D0114651

Atomic Absorption
Spectroscopy

CHEMICAL ANALYSIS

A SERIES OF MONOGRAPHS ON
ANALYTICAL CHEMISTRY AND ITS APPLICATIONS

VOLUME 25

Second Edition

A WILEY-INTERSCIENCE PUBLICATION

JOHN WILEY & SONS

New York / Chichester / Brisbane / Toronto

Atomic Absorption Spectroscopy

Second Edition

MORRIS SLAVIN

Chemistry Department
Brookhaven National Laboratory

A WILEY-INTERSCIENCE PUBLICATION

JOHN WILEY & SONS

New York / Chichester / Brisbane / Toronto

Library of Congress Cataloging in Publication Data

Slavin, Morris, 1901-
    Atomic absorption spectroscopy.

(Chemical analysis; v. 25)
First ed. (1968) written by W. Slavin.
Bibliography: p.
Includes indexes.
    1. Atomic absorption spectroscopy.   I. Slavin,
Walter. Atomic absorption spectroscopy.
II. Title.   III. Series.

QD96.A8S55   1978   543'.085   78-16257
ISBN 0-471-79652-2

Printed in the United States of America

10 9 8 7 6 5 4 3 2 1

FOREWORD

Providing a foreword for this book has proved difficult for me. Partly this is because of my unfamiliarity with the role of commenting on work accomplished by my father. Partly it is because of the character of the book itself.

This is an opportunity for me to acknowledge my indebtedness to my father for stimulating my lifelong interest in the application of physics to the service of analytical chemistry, particularly to the solution of routine analytical problems. My interest started in high school or earlier, as I watched the brilliant colors projected from the electric arc on the laboratory wall while monitoring the arc gap of the spectrograph. This fascination with analysis, especially through spectroscopy, has provided me with endless pleasure and satisfaction.

This book really should not be considered a new edition so much as a new book on atomic absorption analysis from a different perspective. While my wife Sabina and I have discussed many portions of the book with my father, in the end it remains his concept. We hope many workers will find that it facilitates the application of atomic absorption to their practical problems.

Since the original book on atomic absorption was published in 1968, the field has matured considerably. New, broadly applicable techniques appear less frequently. The reliability of analytical results has increased and recent trends in instrumentation provide analytical results more rapidly, with less operator attention and with greater assurance that the correct result will be reported.

Those whose lives have been made easier by atomic absorption owe a great debt of gratitude to the many research laboratories from which the important ideas and techniques have come. Particularly, Sir Alan Walsh has continued to exert a deeply insightful, paternal guidance that has proved to be of considerable value to all of us associated with the field. Professor B. V. L'Vov has exerted the same kind of influence in work related to the furnace, which by now has become a major portion of analytical atomic absorption spectroscopy. When names are mentioned specifically, Professor C. Th. J. Alkemade's name must be included. He

v

was the simultaneous inventor of atomic absorption and has contributed immensely to the understanding of the physical chemistry within the flame.

And finally, as I did in the first edition, I should like to draw attention to the future. Perhaps the greatest technological complement to atomic absorption spectroscopy in the next decade will be provided by plasma emission techniques. The very high temperatures of these sources provide a favorable environment for many of the matrices and elements that have proved difficult by atomic absorption spectroscopy. And, of course, emission methods lend themselves to multielement analysis.

Zeeman spectroscopy is currently being exploited to provide improved correction for nonatomic background absorption. This will have a particularly important effect on furnace methods.

WALTER SLAVIN

Ridgefield, Connecticut
September 1978

PREFACE

The first edition of *Atomic Absorption Spectroscopy* appeared in 1968 and was written by my son Walter. He has since become so occupied with other fields that he has not the time for the work of revision, so both he and the publisher asked me to undertake the job, the reverse of the more usual procedure of the son following the father. I hope that the present edition will find the same favor among workers in the field as did his, which has not only sold well in its English edition but has also been translated into Japanese, Russian, and Italian.

In its general organization this edition follows the first, but there is greater emphasis on the chemistry of the atomic absorption technique, the section on its history has been expanded, and, of course, a section is now devoted to the applications of the furnace as the volatilizing medium of the sample. In 1968 L'Vov's proposal of a furnace in place of a flame was still a subject of somewhat skeptical discussion at meetings of scientific societies. His design of the initial furnace was not particularly attractive, for it required two separate and simultaneous sources of heat—a low voltage current for the resistance heating of the cuvette and an arc, difficult to control, to vaporize the sample. Now, with the availability of well-designed commercial equipment, the furnace is probably the equal of the flame in frequency of use.

The aim of the present book, however, has not changed. It is still addressed to the practical worker at the bench, whom I think of as a person interested primarily in analytical chemistry and only marginally in atomic and optical theory. He or she can easily pursue the theoretical side of the subject through many excellent texts, a few of which I have included in the reference list at the back of the book.

A new feature is the collection, in Chapter 5, of miscellaneous data, physical and chemical, on all the elements amenable to determination by atomic absorption. The data consist, for each element, of its most sensitive line and the lowest concentration of the element for easy determination, together with the linear range to be expected. In addition, secondary lines are listed. For convenience in standards preparation, the preferred compound of the element is stated, together with its chemical formula and with the weight needed to make up a solution concentration that is suitable for storage. All these data are arranged in compact form for ready reference.

In the course of preparing this book for publication, I have consulted closely with my son and with my daughter-in-law Sabina Sprague Slavin, who is equally well known among workers in the field, through both her publications and her close association with the *Atomic Absorption Newsletter*. While on the subject of acknowledgments, I wish to thank the Perkin-Elmer Corporation and the firm of Instrumentation Laboratory, Inc. for their permission to reproduce matter from their publications.

In my search of the literature, I have been struck by the nearly total absence of papers on the problem and application of determining major constituents in the sample, as compared to minor constituents and traces. I could find, among the thousand or so papers, no more than a half-dozen whose authors were concerned with the main problem in this area, the need for high precision. We have apparently missed the golden opportunity of displacing gravimetric analysis by atomic absorption, whose practitioners seem to be fascinated by the ease of treating lower and lower concentrations, forgetting that workers in emission and mass spectroscopy, and in neutron activation, have already done the job, although with more complex equipment.

Atomic absorption equipment is now so stable that, as the few papers extant have shown, precision equal to gravimetric methods, even to so-called referee analyses, is possible. The saving in time and labor can be very great. Moreover, despite the common wisdom, not all gravimetric methods are so exact, as my own experience with one type revealed.

During World War II, the U.S. Government was engaged in buying, for the war effort, tantalite concentrates in Brazil. I was appointed as its representative and a Brazilian chemist represented the seller. Our job was to assay the various shipments.

The best method for tantalum available at that time was the Schoeller tannin procedure, which involved separating the tantalum by repeated differential precipitation of the yellow tannate from niobium and titanium tannates, which were both red. Our initial attempts required fifteen days for the assay and agreement between duplicates was poor. After a good deal of experimentation, we were able to reduce this time to five days by modification of Schoeller's procedure. Only three or four samples could be run in duplicate, and almost invariably we experienced such accidents as a boilover or a broken beaker. We could not determine the standard deviation by doing 10 to 20 repeat analyses on the same sample because of lack of time, but had to rely only on the agreement between duplicates. These agreed only to about 1 part in 50, in the 50% range of Ta_2O_5.

Admittedly, the determination of tantalum in mineral concentrates is an extreme example of a difficult analysis, but it does indicate the direction to follow for a vast improvement by atomic absorption over gravimetric techniques. In the first place, separation of elements with closely similar chemical properties is done surely and automatically by spectroscopy. Second, elements of low sensitivity present no drawback because we are dealing with major constituents, and in general sample quantity is unlimited.

The major constituent field by atomic absorption analysis is well worth cultivating.

MORRIS SLAVIN

Setauket, New York
August 1978

CONTENTS

Go, wondrous creature! mount where Science guides;
Go, measure earth, weigh air, and state the tides;
Instruct the planets in what orbs to run;
Correct old Time, and regulate the sun.

An Essay on Man
POPE

1

HISTORY AND SCOPE OF THE METHOD

1.1 EARLY HISTORY OF THE UNDERSTANDING OF LIGHT

The invention of printing by movable type, in the mid-1400s, caused a proliferation of books and introduced people to the pleasure of reading. But as one enters middle age, physiological changes in eye adaptation makes reading at normal distance difficult (the arms become too short, as the wry saying goes) so, among those with the ability to pay, a demand arose for spectacles. Singer (1) remarks that by the year 1500 spectacles had become common throughout Europe.

The trade of preparing glass and grinding lenses also produced glass prisms as a sideline. These apparently were used only as toys, to amuse people by letting them observe the play of colors through the prism.

Since ancient times the nature of light, and particularly of color, had been a source of curiosity to those educated men who had the capacity to be curious. The Aristotelian view, held from Greek times up to the seventeenth century, was that colors consisted of a mixture of white light and darkness and could be changed by changing the mixture. How confused the ideas of the time were can be shown by citing one of the more respectable theories. Dr. Isaac Barrow (1630–1677), the first Lucasian professor of mathematics at Cambridge University, held forth on the subject of color as follows [quoted by Sawyer (2)]:

White is that which discharges a copious light equally clear in every direction. Black is that which does not emit light at all, or which does it very sparingly. Red is that which emits a light more clear than usual but interrupted by shady interstices. Blue is that which discharges a rarified light, as in bodies which consist of white and black particles arranged alternately. Green is nearly allied to blue. Yellow is a mixture of much white and a little red; and purple consists of a great deal of blue mixed with a small portion of red. The blue color of the sea arises from the whiteness of the salt it contains, mixed with the blackness of the pure water in which the salt is dissolved; and the blueness of the shadows of bodies, seen at the same time by candle and daylight, arises from the whiteness of the paper mixed with the faint light of blackness of twilight.

1

All this was radically changed by the work of that other Isaac, Newton (1642–1727). He had been a pupil of Barrow's, and was his successor as professor of mathematics when Barrow had to flee England during the Cromwellian rebellion because of his royalist sympathies.

In 1666, when he was a young man of 24, Newton bought a glass prism "to try therewith the celebrated phenomena of colors" and started the researches that laid the basis for modern spectroscopy, one of his lesser contributions to science!

His experiments in optics have been often described (1). A ray of sunlight was caused to pass through a hole in a "window-shut" into a darkened room and onto a screen. The prism, placed in the beam, dispersed the light into a spectrum (his term) of colors ranging in the order red, yellow, green, blue, and violet. The refringences of the various colors were measured and compared to the requirements of Snell's law, which had been published a few years earlier. Refracting the individual colors through a second prism, Newton found no change in color. Mixing two rays produced a third color, which could then be analyzed into the two original colors with another prism.

Newton concluded that white light is a "confused aggregate of rays imbued with all sorts of colors, as they are promiscuously darted from the various parts of luminous bodies," and that the colors cannot be changed; hence they could not be considered qualities of bodies, but of the light itself. The function of the prism is merely to analyze the light into its component colors. It is interesting to speculate on the cause of his failure to note the irregularities in the sun's spectrum—what we now call the Fraunhofer lines (3). The aperture of the beam in the window shade was probably a round hole, which would obscure these lines, even though in the later experiments a lens was used to form a sharper image of the spectrum on the screen. If this aperture had been slit-shaped, as we know from hindsight, the absorption lines would have been revealed, and the whole course of spectroscopy might have been advanced by 150 years.

Newton abhorred speculative theories, but his view of the nature of light was that it is a stream of material particles emitted by a luminous body that become visible when they impinge on the eye. (We now call this the corpuscular theory.) He also came close to anticipating Young's development more than a hundred years later of the rival undulatory, or wave, theory. He studied the nature of the colored rings (Newton's rings) formed when the convex surface of a lens is placed in contact with a sheet of glass. Christian Huygens (1629–1695), a contemporary with whom Newton corresponded, proposed in 1678 a wave theory of light that might have explained the formation of these rings. Newton rejected this wave theory because light composed of waves would bend around an obstruction into the shadow, whereas it obviously was propagated linearly. Thus he missed the opportunity to discover the phenomena of diffraction and interference.

Newton's contributions not only mark the beginnings of spectroscopy, but present a fascinating picture of a brilliant mind at work. His experiments were simple and direct; he had few tools with which to work, but nevertheless he drew the correct inferences, always endeavoring to place his observations on a quantitative basis. For the times, this was a momentous change in treating scientific observations, and although others before him, notably Galileo. did the same, none carried this procedure out so completely.

1.2   THE NINETEENTH CENTURY: THE BEGINNINGS OF CHEMICAL SPECTROSCOPY

Nothing of note occurred after Newton's death in 1727 until 1800, when the British astronomer W. Herschel, scanning the solar spectrum with a thermometer, discovered the infrared portion. Soon after, the ultraviolet was discovered by the effect of these rays on silver chloride.

In 1802, Wollaston (4) reported seeing dark lines in the sun's spectrum, but he could give no satisfactory explanation for them. Fifteen years later, Joseph Fraunhofer (5) (1787–1826), with improved equipment, noted the same lines and proceeded to make an extended study of them.

An optician and instrument maker by trade, although he had had no formal education, Fraunhofer improved the method of viewing by placing a telescope after the dispersing prism, which produced a significantly better image. His first spectroscope (Fig. 1.1) had no slit. When a slit was added in subsequent instruments, workers began to call the images they saw "lines."

Fraunhofer also grasped the significance of Young's work on interference in light rays, and Fresnel's on diffraction, which led to his making of the first diffraction gratings. His early gratings were merely crude arrays of parallel wires wound in the threads of two screws, but he went on to

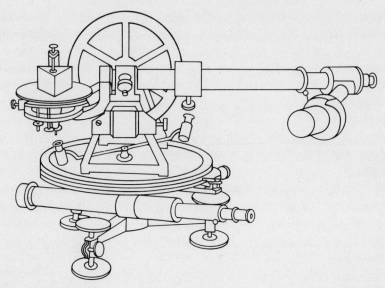

Fig. 1.1 One of the first of Fraunhofer's spectroscopes—really a spectrometer. The divided circle with magnifiers was used for angular measurements, although there was no slit.

make a ruling engine with which he was able to produce really fine gratings of 300 to 600 grooves per centimeter. These were on glass and were used in transmission spectroscopy.

With this equipment, Fraunhofer proceeded to map the dark lines of the solar spectrum, designating the more prominent by letters of the alphabet—a notation we use to this day—such as the D lines for the yellow sodium doublet.

Spectra produced by a grating are nearly linear with wavelength, not like prism spectra in which wavelength is a complex function of the prism's dispersion. This linear relation enabled Fraunhofer to calculate the wavelengths of the dark lines by the geometric relationship between the angle of dispersion and the grating spacing. His measurement of the D lines of sodium (unresolved) gave an average of 0.0005887 mm; our present value of the midpoint of the doublet is 0.0005892 mm. His error, if it may be called that, was only 5 parts in 6000!

Fraunhofer's contributions to spectroscopy were not his only accomplishments. He also showed the world how to make fine glass of specific refractive index (previously it was a hit-or-miss affair) and how to make excellent achromatic lenses and fine refracting telescopes for astronomical use. This astonishing man died at the age of 39. A biographer listed his qualities as "profound perspicacity, powerful inventive genius, tireless industry, strict love of truth, and technical mastery." Contemplating Fraunhofer's work, I am reminded of the words of one of my colleagues at the Brookhaven National Laboratory, who was engaged in the design and construction of the Synchrotron: "A good machinist is worth ten lousy Ph.D.'s."

During the first half of the nineteenth century, a good deal of experimentation took place with the colored flames produced by injecting various salts into a flame. When observed through a spectroscope, bright, discrete lines were seen against a dark background, the reverse of the solar spectrum. For various reasons, the connection between the two was not made for many years.

One of the experimenters was Robert Bunsen (1811–1899), professor of chemistry at the University of Heidelberg. Like the others, he used an alcohol burner or oil lamp for generating the flame, and found them very unsatisfactory. In 1855 piped coal gas came to Heidelberg, and Bunsen at once designed an efficient burner for the new fuel. The burner's flame was much steadier than that of the alcohol lamp, gave a much higher temperature, and was free of impurities. The salts could be introduced into it by means of a platinum wire, and the colors could thus be studied much more conveniently.

It was already known that certain salts exhibited certain colored flames, which indicated that this could be a means of identifying salts. However, when Bunsen viewed the flame through a spectroscope, he noted that the colors were linked to the element, not the compound in which it was bound. It was no great step from there to realize that the bright lines seen with the spectroscope were characteristic of specific elements; here was an extremely sensitive and simple method of element identification. Emission chemical spectroscopy was born.

Bunsen demonstrated the power of the new method by discovering and isolating two hitherto unknown elements in mineral water, cesium and rubidium. In rapid succession, other workers reported the discovery of thallium, indium, gallium, and germanium, all elements whose bright lines are in the visible region. Now, its spectrum was seen to be the means not only of identifying an element but of defining it.

The professor of physics at Heidelberg, G. R. Kirchhoff (1824–1887), became interested in Bunsen's experiments and there followed a close collaboration (6) between the two. Kirchhoff remeasured the wavelengths of many of the Fraunhofer lines, compared them to the line measurements obtained in the laboratory, and showed that they arose from the same elements; thus, these elements were present in the sun. He explained the reversed appearance of the lines as due to a process of absorption as the emission rays passed through the cool outer layer of the sun's atmosphere, which caused them to show up dark against the bright background. Hence, the absorption spectrum is just as characteristic of a specific element as its emission spectrum.

Another conclusion, practically forced on Kirchhoff because of his comparison of solar and laboratory spectra, was that absorption occurs only at the same wavelength as emission; for all other wavelengths the luminous gas is transparent. We call this phenomenon "resonance."

By the final quarter of the nineteenth century, the basis for chemical analysis by means of spectra was firmly laid through the successive accomplishments of Newton, Fraunhofer, Bunsen, Kirchhoff, and their associates. A measure of the advance in the study of the "phenomena of colors" can be gained from a comparison of their conclusions with the views of Isaac Barrow. The explanation of line spectra and their production was not yet known; it had to await the development of the quantum theory in the succeeding century. But knowledge of the quantum theory was not really essential for practical applications.

1.3 EARLY DEVELOPMENTS IN THE ATOMIC ABSORPTION FIELD

From the time of Bunsen and Kirchhoff's demonstration of spectroscopy's effectiveness for elemental chemical analysis, the technique came into general use as a qualitative method, first by visual observation with a spectroscope and later by photography. Modern quantitative methods, aside from the work of a few academic researchers, did not begin until about 1925, when good commercial optical equipment began to appear in the industrialized European countries and in the United States. The method used was emission spectroscopy.

The fundamental process of absorption was well understood, particularly from the work of Kirchhoff, which had explained the dark lines of the solar spectrum. Researchers studied the process in the laboratory, as evidenced from Fig. 1.2, which is taken from a review of nineteenth century work on light and spectra that was published in 1898 (7).

In the early years of the twentieth century, R. W. Wood (8,9) performed the definitive experiments that elucidated atomic absorption by

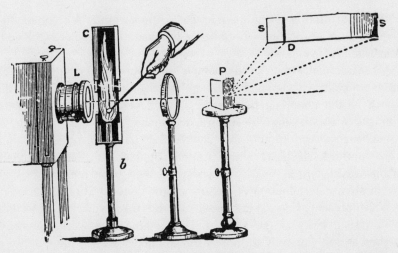

Fig. 1.2   An early arrangement for the observation of absorption spectra. Light from the sun passes through a flame colored yellow by a sodium salt, absorbing that part of the sun's continuum at the position of the Na D lines.

resonance in gases. Soon after his work became known, demonstrations of the mechanism became common in physics lectures in colleges. This consisted in irradiating the fumes above a heated dish of mercury with a mercury discharge lamp, and demonstrating the shadow formed on a screen by the interruption of the lamp beam. In 1939 Woodson (10) applied the procedure to a quantitative method for the determination of mercury; it is a standard procedure to the present day, with minor changes. Woodson's paper appears to be the first application of absorption to quantitative elemental measurement.

Atomic absorption spectroscopy (AAS) as we know it today had its birth in 1955, when two independently published papers both described the method, although in somewhat different terms. One paper was by Alkemade and Milatz (11), working in Holland; the other was by Walsh (12), a staff member of the Commonwealth Scientific and Industrial Research Organization (C.S.I.R.O.) in Australia.

The scheme proposed in both papers was to inject a solution containing the analyte element into a flame, scan the flame with a lamp or other source emitting the spectrum of the same element, and then measure the absorption as a ratio of the intensity of the scanning beam to that of the unabsorbed beam. Both papers pointed out that this absorption is a measure of the atomic concentration in the flame gas, which is a function of the concentration of unknown in the solution sample. A comparison of this degree of absorption with a series of standards completes the determination.

Alkemade and Milatz used a simple filter photometer with a double-beam optical train, chopped the scanning beam, and tuned the recording device to the chopping frequency. They depended on the resonance principle to obtain very high spectral discrimination, thus overcoming the crude spectral separation of their interference filter. Advantages claimed were "optimal selectivity, simplicity, high luminescence and invariability of the wavelength setting." Experimental data were presented on the quantitative determination of sodium in low concentration.

Walsh presented no experimental data, but discussed very thoroughly the physical basis of the method. While the Dutch workers saw the method as applying only to those elements that were usually determined by emission flame photometry, Walsh realized its universality, its applicability to all elements that could be vaporized.

Alkemade and Milatz made no further effort to popularize the absorption method, but Walsh pursued that objective energetically. He describes the early years in a recent publication (13) titled "Atomic Absorption Spectroscopy, Stagnant or Pregnant?" He applied for a patent, exhibited the equipment he had used to test the procedure at a meeting at the University of Melbourne, and endeavored, without success, to interest several instrument manufacturers in commercial exploitation. Even after the appearance of several papers by Australian workers (14–17) who gave their account of the method in practice, there was still no general interest.

Walsh recalls one of his "great moments" in 1962, during this discouraging time:

when I was describing to various staff members of the Perkin-Elmer Corp. in Norwalk the impressive results which were being obtained by the laboratories in Australia which were by that time successfully using the technique. It was during these discussions that Chester Nimitz asked, rather tersely: "If this goddam technique is as useful as you say it is, why isn't it being used right here in the United States?" My reply, which my friends in Norwalk have never allowed me to forget, was to the effect that he would have to face up to the fact that, in many ways, the United States was an undeveloped country!

Soon after this incident several manufacturers brought out complete, integrated AAS instruments and with their vastly greater resources of promotion succeeded in spreading the method throughout the world. The year 1962 should be marked as the real beginning.

An interesting speculation is, Why did the adoption of AAS take so long? All the basic facts and necessary instrumentation were available in 1925 when analytical emission spectroscopy began, with photographic photometry as the means of measurement. But this could just as well have been used for AAS, as Dr. Walsh pointed out to me in a private communication in June, 1977. Only when instruments could be bought as a complete package, followed by a body of literature on the new method, did the great majority of analytical chemists become convinced of the advantages of AAS. Evidently, instruments and literature are the seminal need.

The magnitude of the accomplishment, and the revolutionary effect on the analytical chemistry of the elements resulting from the work of Alkemade and Milatz, and of Walsh, can best be indicated in statistical terms. The number of instrument manufacturers who have entered the field, beginning with Hilger and Watts in England and the original two in the United States (Perkin-Elmer and Jarrell-Ash), has grown to more than 10, with some producing several models. The number of instruments in use today cannot be estimated because many laboratories have assembled their own instruments from separate parts and left no record, but the C.S.I.R.O. licenses under their patent indicates that about 2800 were in use to 1970, when the patent expired (13).

According to the data in the comprehensive bibliography published in the *Atomic Absorption Newsletter,* there were over 5500 publications on the AAS method in the periodical press between 1955 and 1977. Textbooks on the subject occupy a respectable amount of shelf space in libraries. The rapidity of this development is astounding.

1.4   ADVANTAGES AND SCOPE OF THE AAS METHOD

Present-day practice has followed the original suggestions contained in the two pioneer papers. A chopped exciting beam from a hollow-cathode lamp, emitting the spectrum of the analyte, scans a flame into which a solution of the analyte has been injected. The beam, now weakened by absorption by the analyte atoms in the flame, passes to a monochromator, which isolates a single line of the analyte spectrum. A photocell, amplifier, and readout device then measure the intensity of the analyte line. An additional degree of isolation is achieved by tuning the amplifier frequency to the chopping rate, thus preventing the unwanted flame radiation from augmenting the signal. A signal obtained in the same manner but with pure solvent only (no absorption) permits the calculation of percent absorption, which is an indication of analyte concentration in the solution.

The arrangement of the various parts of the instrument is shown in block diagram in Fig. 1.3.

For efficient absorption, the scanning beam must emit a spectrum of sharp lines, and for this purpose both the hollow-cathode lamp and the electrodeless (high frequency) discharge tube are eminently suitable. The burner, which has been studied and improved over the years, should produce a stable flame, to cause a minimum of noise in the signal. The monochromator need be capable of only moderate resolution. The photocell is a photomultiplier, providing more than adequate sensitivity. The readout device can be a simple microammeter, a strip-chart recorder, or, in recent years, a digital readout. Hollow-cathode lamps have also been greatly improved. They are now permanently sealed and are made by several manufacturers for all the elements that can be determined by AAS.

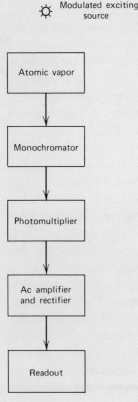

Fig. 1.3 An AAS block diagram of the light path, from the exciting lamp to the final readout.

The components of the optical train are not new. Hollow-cathode lamps have been in use since 1916, and electrodeless discharge tubes have been available for years. Burners and techniques for injecting sample solution into a flame were developed early in the century. Lundegardh, for example, in a series of publications (18) starting in 1928, reported on the solution method for the determination of some 35 metals (this was by emission, not absorption). The choice of fuel has been expanded from the acetylene used by Lundegardh to include nitrous oxide and several other gas mixtures. In place of the flame, the chemist now has available a flameless, high-temperature furnace for the volatilization step. All these devices will be discussed more fully in the chapters on instrumentation and technique.

The AAS method of elemental analysis now pervades every field of chemistry, from geology to biology and medicine. Compared to competing methods, AAS has high sensitivity and high precision, and it can be applied to some 70 elements. The only ones that are not dealt with are the noble gases and several of the common gases whose resonance (most sensitive) lines fall in the vacuum ultraviolet, for which commercial equipment is not designed. Certain of the rare radioactive elements, for which no hollow-cathode tubes are made, also are not usually determined by AAS. On the other hand, common nonmetallic elements, such as sulfur, phosphorus, and the halides, can be treated by indirect methods, with many references in the literature. The operating procedure is simple and fast, and sample preparation in most cases requires a minimum of chemical manipulation. Interferences, which present such severe problems in emission analysis, are at a minimum. The sample weight required is small, and consequently limited samples are no drawback.

In specific terms, sensitivity for the various elements ranges from about 20 μg/ml down to about 0.005 mg/ml and no more than about 1 ml is needed for a determination. This applies to the flame technique; if volatilization is by the electric furnace, which requires only about 5 to 100 μl of solution, sensitivity increases by 1000 times. This is much better than emission and X-ray fluorescence, and is directly competitive with mass spectroscopy and neutron activation.

With regard to precision, the showing is, if anything, more impressive. It is no trick to obtain a precision of 1% relative standard deviation (RSD) with the flame. With the furnace, precision is not quite so good. This applies to routine work; with more care and labor the precision can be improved to about a 0.3% RSD, competing directly with classical referee procedures for the determination of major constituents in a sample. Kahn

(19) has discussed the factors governing high precision analysis, and Price (20) presents certain practical results in the analysis of some common alloys. A recent paper by Fernandez and Korber (21) describes the results obtained by high precision techniques on several Bureau of Standards samples. (See Section 4.1.4, High Precision Analysis.) Errors were no larger than those obtained by the chemists who took part in the establishment of the certificate values.

The operating procedure, once the sample is in solution and made up to standard volume, is simple and rapid. Throughput of samples by the flame technique is about four per minute; for the furnace technique it is slower, about two to three minutes per run. Furthermore, an inexperienced operator can be trained in short order; the more difficult job of sample preparation and organization of the work may require the services of a professional chemist.

Cost of equipment can be low, lower than for most instrumental methods. A basic low-cost instrument is capable of performance near the equal of the most elaborate, although without certain automatic features that add a good deal of convenience. These additions include small computers, teletypewriters, automatic control of furnace temperatures, and automatic sample changers. If the laboratory interest is in many elements, the necessary library of hollow-cathode and electrodeless lamps can amount to a significant sum.

A basic instrument with none of the amenities is cheap enough (about $4000) to be used for the determination of a single element at the point of origin of the sample—as a control in an industrial process, for example. Another possible use for such an instrument is in field surveys, where it could be mounted in a van or truck; power requirements are low and the necessary gases can be carried in steel bottles. Other uses in dispersed locations should readily occur to prospective users. Elwell (22) described the alternative system used by his firm, Imperial Metal Industries, in their British works: "IMI operates two Sendzimer mills, six vacuum-melting furnaces, three atomic-absorption spectrophotometers, and a boy on a bicycle."

AAS is primarily a method for solutions, and this fact adds a great deal to its attractiveness. The prime advantage of solutions is that they are homogeneous, or can be made so. Contrast this with powdered or granular samples, or especially with metal rods such as are used in steel analysis with the direct reader, where a tiny portion of the metallic surface is sampled by the spark. Compared to solutions, inhomogeneity due to segregation presents severe problems for all solid samples.

In any method involving comparison of an unknown to synthetic standards, a basic requirement is that gross composition of the latter must match the unknown, in order to avoid interferences. The composition of complex mixtures would then have to be known, or the major constituents determined, even though they are of no interest. With solutions, the unknown and standards can be matched easily by a process called standard additions, in which the sample carrying the unknown is doped with known additions of the analyte (a term meaning the element of interest) and the unknown's concentration established by extrapolation of the resulting curve. This process is impractical with solid samples.

1.5 SOME LIMITATIONS OF THE AAS METHOD

Things the AAS method does badly or inefficiently or cannot do at all should be known and understood by the analyst. Measurements are made by a single-output channel, so determinations must be made serially. It is true that multichannel measurements can be made, as in direct-reader work, with a continuum source for excitation in place of the individual hollow-cathode lamps. But this would require a high-resolution monochromator and cause other problems that would make the procedure far from simple. For these reasons the multichannel system is very seldom used.

Changing the serial equipment setup from one element to another requires changing the exciting lamp, adjusting the wavelength setting, maximizing of the signal, and making one or two calibrations. None of these requires much time or effort, although lamp warmup for a single-beam illuminating system may take half an hour. However, some single-beam instruments can be equipped with a turret arrangement carrying several lamps in a warmup condition.

Where the AAS method falls down completely is in purely qualitative analysis, for which the large emission spectrograph with carbon arc and photographic recording is without peer. Conventional wisdom, in both industrial and academic organizations, holds that the proper place for qualitative analysis is the spectroscopic laboratory, so that is where such samples are sent. But if the laboratory has no spectrograph, the work will have to be done by wet methods, and this is a vanishing art. Admittedly, a simple inquiry, such as "Is element X present?" can be readily answered by AAS if the appropriate lamp is at hand, or if not, then by use of the emission mode of the instrument, although this will miss several important elements whose sensitive lines are not emitted by a flame. But how cope with a G.K.W. (God knows what) sample?

A questionable procedure sometimes suggested in the literature—a procedure borrowed bodily from the emission field—is the use of an internal standard. This requires the addition of a different element to the unknown solution, followed by the determination of the ratio between the added element and the unknown, either serially or simultaneously using two channels. The theoretical basis for the internal standard is very doubtful, even for emission spectroscopy; besides, two-channel instruments are rare. Furthermore, the scheme is entirely unnecessary, because precision by conventional means is excellent.

ATOMIC PROCESSES

2.1 THE QUANTUM THEORY

Spectra, from their general appearance, can be classified into three types—a continuum in which all wavelengths occur in an unbroken array and arise from glowing solids and liquids; a complex banded or fluted system of lines that are characteristic of bound atoms (molecules and radicals); and single spaced lines, caused by free atoms in a hot gas. It was this last type that mainly interested early investigators, because they felt that study of atoms in a gas, uninfluenced by close neighbors, would lead to a better understanding of atomic structure.

Thus, in the 1880s and 1890s, investigators sought to find some general law that would give meaning to the observed line groupings. Success did not come until it was realized that lines should be denoted by the number of waves per centimeter (the wavenumber) rather than by wavelength. It was found that the lines in the spectrum of hydrogen could be grouped into four series and that a simple expression could cover all four.

For example, in the second series, falling in the visible and near-ultraviolet, the expression takes the form

$$\frac{1}{\lambda} = R \left(\frac{1}{2^2} - \frac{1}{n^2} \right) \text{ cm}^{-1}$$

where λ is the wavelength, n an integer greater than 2, and R a constant whose value was found to be 109,678 cm^{-1} (known as the Rydberg constant). Thus, all lines of hydrogen could be accounted for.

Soon after, Planck, studying the interaction of radiation with matter, could account for his experimental results only by a formula that indicated that energy changes between matter and radiation took place in discrete units, or quanta, not continuously, as was the common view at the time. Furthermore, he showed that a single quantum was equal to the energy frequency times a constant now known as Planck's constant.

Rutherford and his students, working on the problem of atomic struc-

ture, concluded that an atom consisted of a heavy central nucleus sur-
rounded by a number of electrons moving in circular orbits around the
nucleus—analogous to the solar system with its planets. But this did not
explain how stability within the structure was achieved or how spectra
were emitted.

In 1914, using the clue from the old work on series in spectra—that lines
represent a difference between two levels of some sort—and accepting
Rutherford's view of structure, Niels Bohr offered an explanation that is
still accepted.

His radical assumption was that electron orbits represent stationary
states in the undisturbed atom, for which there is no energy change and
therefore no radiation. Radiation would occur only if the electron moved
from one stationary state to another, and this radiation had to be of a
definite frequency, equal to the difference in energy between the two
states, whose levels were fixed by the atomic structure of the particular
element. He assumed that at any level but the lowest the electron was
unstable and immediately returned to the lowest level, yielding up the
absorbed energy in the form of radiation of specific frequency, or as a
spectrum line. Thus, Bohr's paper gave the first reasonable mechanical
explanation of spectrum line origin and of why these lines are uniquely
characteristic of a specific element.

There is no need to proceed further with the historical account. Initial
hints led to far-reaching advances in our understanding of atomic structure and spectra. Textbooks on atomic physics have much fuller accounts
(23).

The atom is now pictured as consisting of a nucleus in which most of the
mass is concentrated, and one or more electrons lying outside the nucleus. The nucleus is made up of protons, each carrying a unit positive
electrostatic charge, and neutrons having the same mass but carrying no
charge. Masses are built up to form the various elements, starting with the
lightest, hydrogen, and going up to the heaviest, according to the system
of the periodic table.

The proton charge of each atom is exactly balanced by the negative
charge of the electrons, which move in circular or elliptical orbits around
the nucleus, with forces outward exactly balanced by the attraction in-
ward, thus achieving stability.

It is assumed that the various electron orbits are arranged in groups or
shells; the configuration of these shells is such that the first contains one
electron for hydrogen, then two for helium, completing the shell. For the

second shell, lithium contributes an electron, making three altogether, then beryllium and so on up to neon, making 8 electrons in the second shell and two in the first. This closes the second shell. The third and succeeding shells are made up in a similar manner until all the elements are accounted for.

Electrons in the uncompleted shells can bind together to form molecules; they are known as "valence" electrons by chemists. Completed shells, which fall at the positions occupied by the noble gases—helium, neon, argon, krypton, xenon, and radon—have no valence electrons and so cannot form compounds. This arrangement of the elements gives a final rational picture of the periodic table originally suggested a hundred years ago by the Russian Mendeleev solely on empirical grounds.

The stationary states are now better called energy levels. By analysis of the spectra of the various elements, the principal energy levels of nearly all the elements have been worked out and their values published in tabular form (24) so that lines can now be assigned to the correct transitions. An elaborate system has been developed for labeling and locating each electron in even the most complex atom. Selection rules for allowed transitions between levels have been stated.

2.2 UNITS AND SYMBOLS

In AAS we deal with isolated or line spectra, which commonly are specified according to their wavelength. The universally used symbol for wavelength is the lowercase Greek letter lambda (λ). The old unit in which lambda was measured was the angstrom, symbolized by Å and equal to a length of 1×10^{-10} meter. Several years ago the IUPAC (25) recommended that the nanometer, symbolized by nm and equal to 1×10^{-9} meter, be used in the chemical literature. One nanometer, therefore, equals 10 Å. The two cannot easily be confused in AAS literature, because the spectral working range is such that a specified wavelength having three digits to the left of the decimal point must be in nm units. If there are four digits before the decimal point, the wavelength is in Å units.

Another line designation sometimes encountered in the literature is wavenumber, which is the number of wavelengths in one centimeter. Wavenumber is commonly symbolized by cm^{-1}, a confusing combination of unit and symbol but hallowed by historical usage. As the nanometer equals 10^7 cm, the relation to the wavenumber is $1 \times 10^7/\lambda$ nm.

A spectrum line is sometimes specified by its frequency, symbolized by the Greek letter nu (ν), with the meaning of waves per unit time. Since the speed of light (with the symbol c) is nearly 3×10^{10} cm/sec, the frequency of a wavelength in these units is

$$\nu = \frac{c}{\lambda\,\text{nm}} = \frac{3 \times 10^{10}}{10^{-7}\,\lambda\,\text{nm}} \text{ cycles/sec.}$$

Since light is a form of energy, a spectrum line can be specified also in energy units. The commonly used unit is the electron volt (symbol eV), defined as the energy acquired by an electron falling through a potential difference of one volt. The wavelength equivalence in terms of electron volts is

$$\text{eV} = \frac{1240}{\lambda\,\text{nm}}$$

and the wavenumber equivalence is

$$\text{eV} = 8067 \text{ cm}^{-1}$$

These expressions are useful when the choice must be made of an analytical line among several in a spectrum, as the line intensity is a function of its energy.

As an illustration of the use of these conversions, consider the strong blue line of calcium, for which we have the following data:

wavelength	422.7 nm
wavenumber	23,650 cm^{-1}
energy	2.93 eV
frequency	7.08×10^{14} cycles/sec

2.2.1. THE ENERGY LEVEL DIAGRAM

An energy level diagram presents a wealth of information of an atom and its structure. A diagram representing the sodium atom is shown in Fig. 2.1. The vertical scale indicates the energy in electron-volts; often the wavenumber is used. The energy levels are represented by horizontal lines at the appropriate positions, and the vertical arrows represent electronic transitions. The arrows show transitions from a higher to a lower level, indicating that the transition results in emission of energy. An arrow pointing up would indicate an absorption of energy.

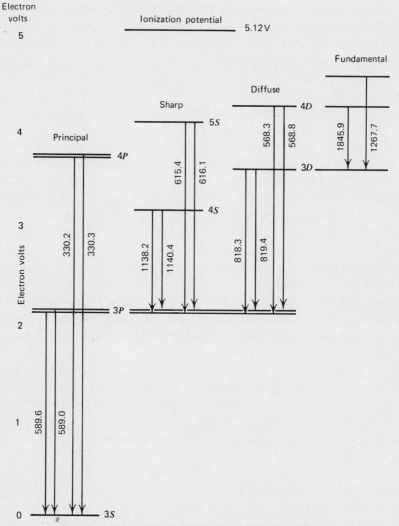

Fig. 2.1 Energy level diagram for the element sodium.

The lowest level is called the ground state; this is the ordinary state in which energy is taken to be zero. Lines terminating in the ground state are called resonance lines; because they are at the lowest level and the easiest (generally) to excite, they are the strongest lines in the spectrum of an atom. In the case of sodium, the resonance lines are the doublet at 589.6 and 589.0 nm (the well-known D lines), of which the 589.6 line is the stronger. Another resonance doublet is the one at 330 nm, but it is much weaker than the D lines.

When an atom's structure contains levels made up of two or more close sublevels, transitions to or from these multiple levels produce multiple spectrum lines of very similar character called multiplets. In the sodium atom, for example, the 3P level is doubled, producing a series of doubled lines, or doublets. In more complex spectra the multiplet may consist of as many as seven separate lines.

Line groups, or spectral series, first discovered by Balmer toward the end of the last century, are marked in the diagram with the terms principal, sharp, diffuse, and fundamental, abbreviated as S, P, D, F, which also serve to label the levels. Transitions, according to certain selection rules, are allowed only between specified levels. The level designations (SPDF) serve also in a system of nomenclature to locate the electron position and to indicate the transition.

It is obvious from the diagram that if an electron is to undergo a transition from a higher to a lower level, it must first be raised to the higher level. This explains the fact that flame and furnace spectra, which are produced in low-temperature sources, are much simpler than those produced in the arc or spark, because the energy available in the former is insufficient to raise the electron beyond the low excitation levels. For example, in the principal series of Fig. 2.1, energy required to raise the electron for the D doublet transition is $1240/589.3 = 2.11$ eV (see Section 2.2), whereas for the 330 nm doublet the energy needed is 3.75 eV. The 3P level represents the first excitation level; it is the lowest for the sodium atom and the easiest to reach, and the transition to and from this level produces the strongest line.

When an electron has absorbed sufficient energy so that it is driven completely out of the influence of the nucleus, the atom is said to be ionized. For sodium, this energy is 5.12 eV. This ionization potential then becomes the ground level for a new and different system of levels similar to that of the element in the periodic table whose outer shell contains one electron less; in our example this would be neon. The un-ionized atom is distinguished by the roman numeral I after the element symbol, the singly ionized ion (which has lost one electron) is designated by roman numeral II, as Na I or Na II, and so forth.

Many other details of modern atomic theory—selection rules, the system of nomenclature identifying each electron in an atom and its location in the various orbits, the energetics involved in collisions with various

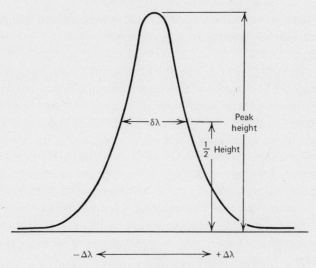

Fig. 2.2   Typical profile of a spectrum line.

particles, and so on—are beyond the scope of this book. The literature on atomic physics and structure is extensive and can be pursued in many excellent texts (23).

Sometimes discussed is the width of spectrum lines, both in emission and in absorption. The profile of a line, plotted as intensity versus wavelength $\pm\Delta\lambda$, follows a gaussian distribution because it is caused by random processes (Fig. 2.2). The peak defines the wavelength, and the width is conventionally measured at half the peak height, called the half-intensity width, $\delta\lambda$. The area under the profile, which is a measure of the total radiant energy of the line, can be approximated by multiplying the peak height by the half-intensity width.

2.3   TRANSFER OF ENERGY BY RESONANCE

The Fraunhofer lines in the solar spectrum, which so interested Kirchhoff, could not be explained by him but had to await the development of modern atomic theory. In the first quarter of the present century, R. W. Wood and his colleagues advanced our knowledge of the mechanism, which is really a special case of interaction of radiation with matter. In 1961 Mitchell and Zemansky (26) devoted a book to the subject.

The interaction is often demonstrated in this way. The mercury spectrum contains a strong line at 253.7 nm. When a beam of light of this wavelength is passed through a mass of mercury vapor at low pressure, it is strongly absorbed, although wavelengths in the spectra of other metals pass through undiminished. The absorbed energy must be accounted for in some way if conservation of energy is to be maintained, and it has been noted that the vapor spontaneously emits the ray 253.7 nm in all directions in the form of a glow. This radiation, to which Wood applied the term resonance fluorescence, is now the basis of an analytical method called atomic fluorescence spectroscopy (AFS). AFS shares both the literature and the instrumentation with AAS, with only minor changes.

The resonance process, which to an observer appears to be simple absorption, is in reality a conversion of a narrow beam into a spherical beam of the same total energy and the same wavelength. The small portion of the fluorescence reaching the observer is so weak as to be negligible.

Explanation for the resonance phenomenon follows directly from quantum theory. The energy absorbed by an atom drives an electron from a lower energy level, generally the ground state, to a higher level equivalent in energy to the absorbed photon (because only "allowed" levels are capable of containing the electron). This excited electron then spontaneously drops back to its original level, and the photon is emitted as resonance fluorescence.

The process has an exact analogy in sound. For example, if two tuning forks tuned to the same frequency are placed a short distance apart and connected only by air, and then one is struck, the other will pick up the vibrations and sound the same note in "sympathy."

2.3.1 MISCELLANEOUS FACTORS PERTAINING TO RESONANCE ABSORPTION

Of some interest are certain miscellaneous factors affecting AAS operation. For efficiency, the fraction of analyte atoms in the absorbing gas that is in the excited state and therefore not available for absorption should be kept as small as possible. Given the statistical weights and temperature and assuming equilibrium, this fraction g_q/g_0 is easily calculated using Boltzmann's distribution law (27). It has been found to be 1% or less at usual flame temperatures. The great mass of atoms are therefore in the ground state and subject to absorption.

Another factor influencing absorption is the requirement that the exciting line from the hollow-cathode lamp be narrower than the line that ordinarily would be emitted from the absorbing gas. The width of a spectrum line depends on both collisional processes and the Doppler effect (28–30), of which the latter is the greater. The Doppler effect is a matter both of temperature and pressure at emission. However, the gas within the lamp is at a lower temperature and pressure than the flame gas, so the line from the lamp must always be the narrower.

The concentration of the analyte injected into the flame has an important influence on absorption by the exciting beam. As it passes outward through cooler layers of its own atoms, radiation originating within the source (the flame) is subject to absorption by these atoms. The gradient of temperature, from the level of emission to room temperature, furnishes a path that contains an aggregate of free atoms just right for this absorption. At low atom concentrations, the line profile is gaussian; at increasing concentration the peak increases, although not in proportion to concentration. The result is illustrated in Fig. 2.3, where curves *1* to *3* still have the expected profile, but curves *4* and *5* exhibit a dip in the peak, with an increase in the spread of the wings. An exciting beam, confined to a narrow range of wavelengths around the peak, then encounters fewer

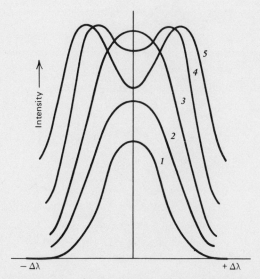

Fig. 2.3 Changes in line profile with increasing analyte concentration. As the concentration increases from *1* to *5*, the profile broadens and peak height increases slower than increase in concentration, until at high concentration the peak actually decreases. From reference 45.

atoms in the absorbing condition, with consequent lowering of the expected absorption.

The phenomenon is called <u>self-absorption,</u> and at very high concentrations the peak disappears entirely, as in Fig. 2.4, with a large increase
in line width. This extreme condition is called self-reversal. The entire
process of self-absorption is well explained by Boumans (31).

Just as with spectrophotometric absorption of molecules in solution,
the analogous process in flames follows the Beer-Lambert law, but only at
low concentrations. Calibration curves are here straight lines passing
through the origin, but at increasing concentrations the slope progres-
sively decreases and can even reverse in direction.

2.4 WAVELENGTH LISTS AND LINE ATLASES

Sources of wavelength information are readily available. The most
comprehensive is the M.I.T. atlas of lines, by Harrison (32), which lists

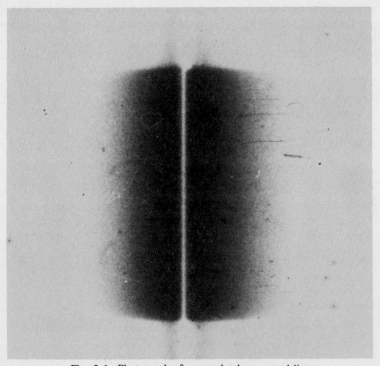

Fig. 2.4 Photograph of a completely reversed line.

over 100,000 lines of some 80 elements, arranged by wavelength and with estimates of line strength derived by visual estimates in photographs. The volume also contains two shorter lists, of the principal (strongest) lines arranged both by element and by wavelength.

A publication by Saidel et al. (33), evidently copied from the M.I.T. publication, contains several lists. The two principal ones are shorter lists from which the weakest lines have been omitted; one list is arranged by wavelength and the other by element. Several much shorter lists in this publication contain data that are not very useful for absorption work.

A serviceable list, arranged according to element, can be found in any edition of the *Handbook of Chemistry and Physics* (34) (the familiar Chemical Rubber *Handbook*).

A fourth source is the two-part set of tables by Meggers, Corliss, and Scribner (35), a publication of the National Bureau of Standards. Volume I is a listing by element and Volume II by wavelength. The data were produced by actual photometric measurement of all line intensities emitted by a copper arc operating at a temperature of about 5000°K. Additional information shown for each element are the atomic number and atomic weight, configurations of the normal state of the valence electrons for both the un-ionized and singly ionized atoms, the ionization potentials, and the wavelengths and wavenumbers of initial and terminal energy levels represented by the transition of each line listed. It is thus easy to pick out the lines whose initial level is at the ground state or at a few wavenumbers higher.

Chapter 5 includes listings of both the best line and some minor lines of each element, taken from the literature or in the Company laboratory and collected by the editors of the Perkin-Elmer "Cookbook."

2.4.1 SELECTION OF THE ANALYTICAL LINE

As AAS is a method mainly for trace determination, the line characteristic chiefly sought is sensitivity, which can be defined as the line's responsiveness to specific element concentrations. This is by no means the only desideratum, however. Other criteria besides theoretical sensitivity are the spectral range of the monochromator; the response of the detector, which becomes poor at either end of the wavelength range; the flame and molecular background; and the line intensity of the exciting lamp, which is in emission, not absorption. Furthermore, in work near the detection limit, it may be better to choose a weaker line in preference to sample dilution.

Much effort has gone into attempts to find a theoretical basis for determining line sensitivity. Meggers et al. (35) define the strongest line in an emission spectrum as the one:

which originates in any spectrum with a simple interchange of a single electron between the S and P states, usually preferring configurations in which only one electron occurs in such states.

The intensity in absorption for ground state lines of an element should be in proportion to the oscillator strength or "f" value associated with the line. These should be directly applicable to atomic absorption, and many workers have found some correlation between the observed strength of absorption lines and the f data. Many of these f values are still to be determined, and accurate numbers for many others are not available, but with continuing work on the problem it is hoped that determinations of many metals will be possible without the need for standards. This subject has been treated in an excellent article by Margoshes (36). In a more recent work, Parsons, Smith, and McElfrish (37) have calculated the theoretically strongest lines in absorption for 32 elements and compared them with the corresponding strongest lines found by experiment. On the whole, the agreement was good. These authors also include a list of atomic populations at the lower energy levels for various temperatures.

It should be noted, before leaving the subject, that the strongest lines of about half the elements are the same for both emission and absorption, and that the latter, almost without exception, are ground state lines. Furthermore, several of the rarer elements have not been thoroughly investigated, either theoretically or experimentally, and therefore some disagreement exists. For a few elements only one line is suitable, and for cerium no line has been found suitable. Errors in line identification have been noted; the most accurate source of wavelengths is the M.I.T. atlas, which should be consulted by authors of publications.

The strongest lines in emission and absorption of an atom are not always the same because the former are caused by particle collisions in the gas and therefore are strongly influenced by temperature, while the latter are caused by resonance energy transfer and are only slightly influenced by temperature. Where the strongest lines are different, the absorption line has the higher frequency or shorter wavelength.

3

INSTRUMENTATION

Walsh, in a recently published article on reminiscences of early experiences with instrumentation, remarks "I would like to think that some of them [instrument makers] are musing on possible ways of embellishment to insure that any commercial version will have an impressive price tag." This thought must have arisen from memories of his difficulties with pickup equipment. The instrument makers have indeed embellished, but they have also assembled the necessary parts and enclosed them in attractive cases and provided convenient controls. They must also be credited with providing a working package to analytical chemists who do not have Walsh's instrumental skills.

The basic equipment, as diagrammed in Fig. 1.3 and discussed in Chapter 2, consists of some sort of scanning beam that irradiates a flame into which a solution of the analyte has been aspirated, and then passes on to a monochromator to isolate a single wavelength, and then to a photocell and readout devices. The separate components of the train will be discussed in detail in this chapter.

A lengthy presentation of basic instrument design has been published by Kahn (38), and more recently the subject has been reviewed by Price (39). A book by Veillon (40) describes all the commercial instruments available up to the date of publication (1972), provides manufacturers' addresses, and lists accessories.

3.1 OPTICAL PATH ARRANGEMENTS

Some early instruments used a single beam and a continuously powered excitation source (Fig. 3.1a). This resulted in a direct current to the meter. With this dc system there was no way to distinguish the exciting ray from a ray of the same wavelength that may have originated from emission in the flame, thus mixing the two. This problem was solved by modulating the exciting source and not the flame. Since the modulated light produces

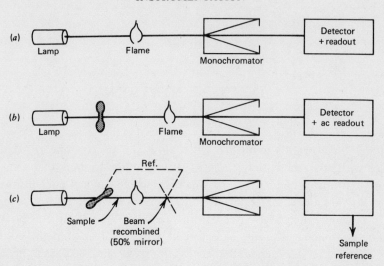

Fig. 3.1 Three optical path arrangements. (*a*) Single-beam dc system; lamp and flame emission both unchopped. (*b*) Single-beam ac system; lamp emission chopped, flame emission unchopped. (*c*) Double-beam ac system; 1 lamp emission chopped, flame emission unchopped. One beam passes through flame, other bypasses flame, but they are later combined and their ratio measured.

an alternating current in the detector, the system is known as a single-beam ac arrangement (Fig. 3.1*b*). The electronics are designed to block any direct current—with a blocking capacitor, for example—from producing a signal, and to tune the amplifier to the frequency of the alternating current. This effectively prevents any flame light from reaching the readout device. The method of modulation is shown schematically as a rotating blade or apertured disc, but the same effect can be achieved by driving the exciting lamp with alternating current.

Another variant is that of Fig. 3.1*c*, in which the exciting beam is split in two. One half of the beam bypasses the flame and is then combined with the analytical beam at the detector to provide a ratio of the two. This is an ac double-beam system.

The advantages and limitations of the double-beam system are worth discussing. Clearly, the double-beam system cannot overcome instability and noise in the burner, since the burner is in only one of the beams. But stable burners are now available; the primary instability in the system is thus the source lamp, which requires a considerable time to come to thermal equilibrium and emission constancy. The double-beam system is largely immune to lamp drift, as shown in Fig. 3.2. The upper figure is a

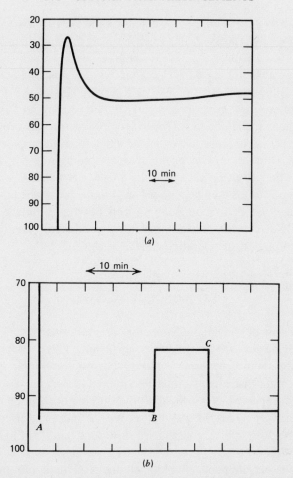

Fig. 3.2 Single- and double-beam operation compared. Hollow-cathode lamp operated from cold start. In (a) with single beam, stability is reached only after about 40 min. In (b) with double beam, baseline is stable immediately. At point B, solution containing calcium is aspirated. At point C, aspiration is stopped.

recorder trace of the output at 422.6 nm of the calcium line from a hollow-cathode lamp in the single-beam mode. Emission still has not come to equilibrium after 60 min. The lower figure shows the output of the same line and the same lamp in the double-beam mode. Almost from the start stability has been reached. At point A the burner is turned on and distilled water is aspirated. At point B a solution containing 1 μg/ml of calcium is added. At point C the solution is removed. Thus, usually, the double-beam system results in a stable baseline almost immediately, with the following advantages: Lamps can be inserted and operated at once with little or no warmup; the baseline stability makes small signals apparent, thereby improving the detection limit over the single-beam mode; and a stable baseline improves precision. Disadvantages of the double-beam arrangement are its greater complexity and the reduction of signal strength reaching the detector. Much work in atomic absorption can be done perfectly adequately with simpler single-beam equipment.

A partial solution to the problem of warmup time for single-beam operation is the provision in some instruments of a rack into which several hollow-cathode tubes may be plugged and connected to the power supply. The waiting time before an analysis can be run is thus considerably shortened. But the life of the lamp is also reduced, since lamp life is basically a function of operating time.

3.2 MONOCHROMATORS

The function of the monochromator, as mentioned previously, is to isolate a single line of the analyte's spectrum. This can be done with filters—better with interference filters—but because their bandpass is very broad, filter instruments must be confined to the analysis of elements having simple spectra. Nevertheless, filter atomic absorption instruments are commercially available and may prove satisfactory for certain problems. Their advantages are simplicity, low cost, and high light transmittance.

The dispersing elements of older commercial monochromators were prisms either of glass or fused quartz, but around 1938 reflection gratings came into the market, to be mounted in large emission spectrographs. Later, a replication process was invented, more efficient ruling engines were built, and gratings became better and relatively inexpensive and thus largely replaced prisms.

The advantages of gratings are that (1) the wavelength range is much broader than that of glass, (2) the reflected beam intensity is much less wavelength dependent than the transmitted beam through the prism, and (3) the dispersion is almost linear, simplifying the wavelength scale.

When a photocell is mounted behind the exit slit, the assembly is called a spectrophotometer. The external illuminating parts are usually mounted on a rigid optical bench securely fastened to the monochromator frame, to ensure that the illuminating beam stays on the optical axis.

3.2.1 OPTICS OF THE MONOCHROMATOR

The internal optics of the monochromator may be arranged in one of several ways. For AAS instruments three arrangements are in common use: the Ebert, Czery-Turner, and Littrow mountings. All three are effective, with very little to choose among them.

The Czerny-Turner mounting, shown schematically in Fig. 3.3, uses two concave spherical mirrors, M_1 and M_2. The external beam enters at slit S_1, is rendered parallel by M_1, and goes on to the plane reflection grating G, where it is dispersed and reflected to M_2. From here the condensed beam goes through the exit slit S_2 and on to a photocell mounted behind the slit. The specific wavelength leaving S_2 is controlled by simple rotation of the grating about its vertical axis, to which is attached a scale calibrated in nanometers. The optics of the grating will be explained in Section 3.2.3.

Slits for simpler instruments are fixed in width and are sometimes mounted in holders that can be changed for an assortment of widths. A much better arrangement, found in the more elaborate instruments, is the use of continuously variable slits, with a scale that indicates the actual width. The jaws move symmetrically; if the adjustment were by one movable jaw, the wavelength setting would shift every time the width was changed.

Modern gratings are ruled with diamond scribers whose shapes are carefully controlled to form grooves that concentrate the reflected beam in a specific direction. This operation is called blazing and its purpose is to increase the light energy into the wavelength region where the most work

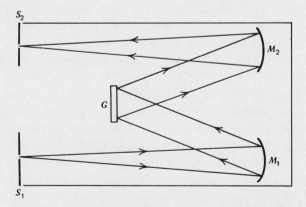

Fig. 3.3 Light path in the Czerny-Turner grating mounting.

will probably be done. The more detailed instruments employ two gratings, so mounted as to be easily interchanged in the optic axis and blazed at approximately 250 and 500 nm. If two gratings are to be provided, this gives the designer the additional option of supplying gratings with two different dispersions, the larger for the ultraviolet, where the spectrum is usually the more crowded.

Based on diffraction theory, a grating reflects not only the wavelength for a specific angle, but also the higher harmonics, a phenomenon exactly analogous to sound. The higher harmonics, called orders of the grating, cause overlapping of rays from totally different spectral regions. For example, lines at 600 nm in the first order, at 300 nm in the second order, and at 200 nm in the third order will all fall on the same point of the focal plane and pass through the exit slit. In practice, this could sometimes cause an obscure and very puzzling interference; this effect should always be borne in mind. Overlapping can always be negated by the use of an absorbing filter in the optical path.

In manufacturers' brochures the optical properties of the monochromator are usually specified as to five particulars: focal length, wavelength coverage, grating area, number of grooves per millimeter, and dispersion. Focal length is the distance between the curved mirror and one of the principal planes when the other plane is at infinity. It is a design consideration, controlling the dispersion and the overall size of the instrument.

Wavelength coverage, as the term implies, is the region of the spectrum that can be observed. In the infrared, the extreme line of interest is 852.1 nm of cesium. Beyond this, there are no lines of analytical use. At the ultraviolet end, on the other hand, there are several valuable lines of common elements, but the air becomes increasingly opaque and this limits the range to about 190 nm. The working range is thus 852 to 190 nm; however, some instruments do not accommodate these extremes because they were designed for analytical work within particular spectral regions.

Grating area is important because it usually is the factor that limits the speed of the monochromator. The grating is the most expensive optical part to produce and so is usually made the diaphragm stop or limiting aperture of the system. It is analogous to the maximum stop or opening of a photographic objective, and the speed is reckoned in the same way—the equivalent diameter of the grating divided by its focal length. Note that grating speed is determined, and therefore stated, assuming that the entire grating face is illuminated but this depends on the system illuminating the entrance slit.

Resolution is a dimensionless number that expresses the ability of an optical system to separate two close lines. Dispersion is the term expressing the monochromator's ability to spread the spectrum along the focal plane. Both resolution and dispersion will be discussed further in a later section.

3.2.2 ILLUMINATION OF THE SLIT

The curved mirror or mirrors within the monochromator comprise the focusing elements by which an image of the entrance slit is formed on the focal plane. This image is at unit magnification; therefore the simplest configuration of entrance and exit slits is to make them alike, both as to width and as to length. Length of the slits must be limited by the diameter of the detector window, which would give the maximum light passage.

As to width, in modern instruments both slits are made to be bilaterally adjustable. Slit width is changed by means of a micrometer screw carrying a scale marked in either millimeters or micrometers. Actual slit width can be expressed in these terms or in spectral bandpass. Bandpass, expressed in nanometers, is determined by multiplying the width of the slit image at the focal plane by the dispersion. For example, for a width of 0.01 mm and a dispersion of 1.6 nm/mm, the bandpass would be 0.016 nm.

The slit width controls, to a degree, certain spectral interferences—unabsorbed lines of the fill gas or unresolved and unabsorbed lines of the cathode material, and molecular bands in the flame. Modulating the exciting beam is supposed to eliminate this interference from the flame or furnace, but it is not entirely effective. These interferences can be controlled, up to a point, by reducing the slit width.

Slit illumination problems are treated by Slavin (41), and general grating theory and arrangement are thoroughly discussed by Barnes and Jarrell (42).

Mechanical considerations impose a lower limit to how much a slit may be closed down. For a good quality slit, this is about 10 μm, which is greater than the width at half-intensity of the sharp lines emitted by a hollow-cathode lamp, so that the bandpass for such lines is effectively equal to their widths.

Radiant power passing through a slit is proportional to the square of the width, but this holds only for a continuum, not monochromatic radiation. As a slit is narrowed, continuum background, or unresolved band structure, becomes less intense, but separate lines are little affected; the ratio of line to background is increased.

If the illuminating system is set up correctly, the collimating system of the monochromator will be filled with light. The collimating lens will be fillled when the angle subtended by the source at the slit is equal to or greater than the angle subtended by the collimator at the slit. If this angular condition is met, no system of condensers will increase the radiant power. In addition, for correctly designed monochromators, the grating will be fully illuminated.

3.2.3   GRATING THEORY

The way a grating "works" can be seen from the following discussion. Figure 3.4 represents two adjacent rulings on the surface of a plane reflection grating illuminated by light containing all wavelengths. The two

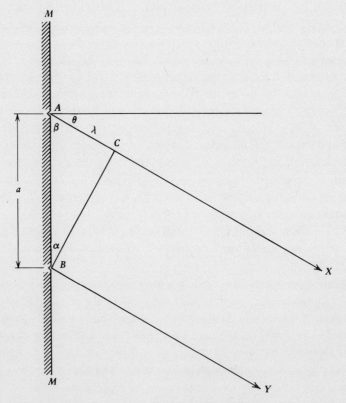

Fig. 3.4 When two light rays x, y pass through adjacent rulings A, B of a grating, such that $AC = \lambda$, they will interfere constructively.

rulings are at A and B, separated by a distance a, the ruling space. Consider two of the reflected rays, moving in parallel paths X and Y, at an angle θ to the mirror normal. A perpendicular BC to the ray X will cut a segment AC on ray X. If $AC = \lambda$, the two rays will be in phase (peaks and hollows coincident) for the wavelength λ and will interfere constructively; that is, their amplitudes will add.

As the grating has thousands of grooves, the reflected rays X and Y must be multiplied by the number of illuminated grooves to arrive at the sum of their amplitudes. The bundle of rays leaving in the direction θ is therefore extremely intense compared to any random light that might take the same direction. To form the actual spectrum line associated with the angle θ it is only necessary to condense the parallel rays to a sharp focus, which is the function of the second curved mirror within the mono-chromator.

This relatively simple action of the grating is the means by which rays from a complex mixture of elements, regardless of whether their properties are similar or diverse, are separated. Compare this process to classic separations by chemical precipitation!

The relation among α, λ, and the groove spacing a is shown to be

$$\sin \alpha = \frac{\lambda}{a}$$

but

$$\alpha + \beta = 90° = \beta + \theta$$

and therefore

$$\sin \theta = \frac{\lambda}{a} \quad \text{or} \quad \lambda = a \sin \theta$$

We have here a relation defining wavelength in terms of the grating spacing and the angle of reflection.

At some greater angle the two parallel rays from two adjacent grooves will be separated by two wavelengths. Reinforcement will again take place, but this time at an angle associated with 2λ (the distance AC in Fig. 3.4 will be 2λ); consequently there will be an overlap of spectra, as explained previously.

In the argument above, it was assumed that the incident beam was normal to the grating surface. For the practical case the beam can be incident at an angle other than normal, in which case the above expression becomes

$$\lambda = a(\sin i + \sin \theta)$$

where i is the angle of incidence. This is the expression generally known as the *grating law*.

In the above equation it was assumed that both incident and reflected beams lie on the same side of the normal; for the case where the two beams lie on opposite sides the smaller angle bears a minus sign.

If the equation is differentiated with respect to λ, incident angle i, which is a constant because slit and grating positions are fixed, drops out and we obtain

$$\frac{d\theta}{d\lambda} = \frac{1}{a \cos \theta}$$

which is the angular dispersion that describes the angular spread of the spectrum along the focal plane.

We are more interested in line separation in the focal plane than in the angular spread. Indeed, manufacturers of spectroscopic equipment specify their instruments in terms of the number of nanometers contained in 1 mm at the focal plane. This figure of merit is called the linear dispersion and, obviously, the smaller the number the better the instrument. To convert angular to linear dispersion, θ must be multiplied by the distance between the condensing medium and the focal plane, which distance is generally but not necessarily the nominal focal length of the optical system. Calling this distance F, and the linear dispersion p, we obtain the dispersion formula

$$\frac{d\lambda}{dp} = \frac{a \cos \theta}{F}$$

This states that the smaller the groove spacing a, the larger the angle λ (smaller $\cos \theta$), and the greater the focal length F, the fewer nanometers within the unit distance p.

With instruments employing gratings as the dispersing medium, one works at small angles θ, so that $\cos \theta$ is nearly unity and changes slowly over the wavelength range, producing a nearly normal spectrum and therefore nearly evenly spaced intervals on the wavelength scale. In older prism instruments the dispersion angle change is much greater, with a consequent uneven scale.

Resolution, or the ability to separate two closely lying lines, has already been mentioned as depending on the total number of grooves in the grating. The conventional way of expressing resolution is by the dimentionless ratio $\lambda/\Delta\lambda$. In words, this ratio is the wavelength interval just separated divided into the wavelength at that point in the range.

The resolution equation as generally written, and without showing the derivation, is

$$R = \frac{\lambda}{\Delta\lambda} = n$$

where $\lambda/\Delta\lambda$ is the interval just resolved and n is the total number of grooves in the grating. It should be pointed out that n is the total number illuminated; if the beam passing through the slit does not go on to illuminate the whole grating, then resolution is reduced.

In practice our interest is in learning whether two specific lines can be resolved; for this the transposed form of the equation is useful

$$\Delta\lambda = \frac{\lambda}{n}$$

3.2.4 THE MULTIPLIER PHOTOTUBE DETECTOR

The old photocell, consisting only of a simple cathode and anode, was of very low sensitivity and required a good deal of external amplification to make signals from low light levels useful. This photocell has now been universally replaced by the multiplier phototube, generally called a photomultiplier or PM tube for short, which contains its own amplification system and requires no external means of increasing the signal.

The internal construction of the PM tube is shown in Fig. 3.5, which is known as a focused dynode arrangement. Other arrangements of the dynodes are used, but the principle is the same. The PM tube consists of a highly evacuated glass vessel with a window, either at its end or at its side, to receive the light beam. Electrical connections are made through the base, which is of plastic, not very different from the ordinary radio tube. The inner wall of the window is coated with a special electron-emitting material, which constitutes the cathode of the electric circuit. A series of secondary cathodes, called dynodes, marked by numbers *1* to *12* in Fig. 3.5, are activated by the electrons emitted from the cathode. The last of these dynodes is electrically connected to the anode, *13*.

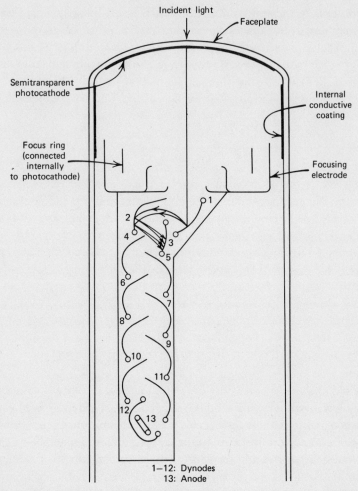

Fig. 3.5 Construction of one type of photomultiplier tube, showing cathode, anode, and dynode array.

Each dynode is connected to the next by a resistor, as shown in Fig. 3.6. A source of dc voltage is applied between cathode and anode. Each dynode, from *1* to *12*, is then at a more positive potential than the previous one.

When light strikes the cathode, the emitted electrons are attracted to the first dynode because of its more positive charge; this causes several secondary electrons to be removed for each primary electron that strikes

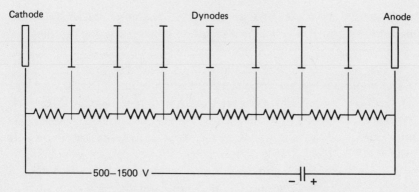

Fig. 3.6 Internal electrical connections of a photomultiplier.

the dynode surface. The process is repeated down the array of dynodes, with each step increasing the number of electrons attracted by the increasing positive charge. This greatly increased current flow to the anode constitutes amplification.

The degree of amplification is easily calculated. If the cathode current due to the primary electrons is represented by N_0 and each of these electrons produce N_s secondary electrons per dynode stage, the total amplification is

$$\text{amplification} = (N_0 \cdot N_s)^n$$

where the exponent n is the number of stages. Commercial PM tubes may have as many as 14 stages, with amplification factors of a million or more.

Amplification levels may be changed by changing the applied voltage, so scale setting and noise levels are within easy control.

Extreme sensitivity is not as necessary in AAS work as in emission photometry, for at low analyte concentrations the light intensity reaching the PM tube is not much lower than the full hollow-cathode lamp intensity. At high analyte concentration this intensity will be greatly reduced, but here certain expedients, such as diluting the solution or turning a slot burner 90° to the light beam, are always available to the operator.

Other advantages of the PM tube for AAS operation are its stability, its linear response over a range of 10^6 times, and its almost instantaneous response time. It does, however, have one failing that is important to AAS workers—it is wavelength sensitive, so it must be selected to fit specific wavelength regions. For many years the standard tube was the RCA IP 28, whose response covered the range 200 to 500 nm, although it tailed off badly at the ultraviolet end. In recent years improvements have been made by American, English, and Japanese tube makers, so that the whole usable spectroscopic range can be covered with no difficulty, although not with the same efficiency over the entire wavelength range.

The photomultiplier's response to radiation requires that the energy of the radiation exceed the electronic work function of the cathode material, so that electrons will be freed from the surface. This, obviously, is a problem only at the long-wavelength end of the spectrum; practically, the problem is with the strongest resonance lines of cesium, rubidium, and potassium at 852, 780, and 766.5 nm. For these lines, the $(Cs)Na_2KSb$ surface with an S-20 resonse is the most suitable, although all three elements have secondary resonance lines that are weaker but fall in a portion of the spectrum whose resonse presents no problem.

Manufacturers of PM tubes issue comprehensive catalogs presenting data not only on the electronic characteristics of their products but also on tube size, socket type, window material, and other physical information. These catalogs are a very good source for the prospective user of photo-receivers. A textbook on the general subject of the photoemissive effect has been published by Sommer (43).

3.2.5 READOUT DEVICES

Absorption in an atomic gas usually follows Beer's law, which states that the concentration of atoms irradiated by an absorbed beam varies linearly with the logarithm of the absorption. If C is the concentration and I_0 and I the beam intensities before and after absorption, respectively, then

$$k\ C\ =\ \log \frac{I}{I_0}$$

where k is a constant of the system. The quantity on the right of the equation is called the absorbance, or sometimes the optical density, although only the former term will be used in this book. For materials that obey Beer's law, the straight line that results from plotting concentration against absorbance is a great convenience, as statistical data on precision can be readily calculated and interpolation in the curve made with a slide rule or desk calculator.

If the readout meter is adjusted so that it reads 100% (full scale) for the exciting beam when only the solvent, containing no analyte, is aspirated, and zero when no light falls on the photocell, then the absorbance A is

$$A = 2 - \log \text{(percent transmission)}$$

Measurements of absorbance can be made with a simple microamme- ter, a potentiometric strip-chart recorder, or a digital readout. The re- corder has been favored in the past, but it is slow in that it requires the tabulation of peak heights from the chart and the handling of long lengths of paper. For routine analysis the meter is much faster. However, the recorder has some advantages: the record is permanent and objective, and noise in the signal is at once apparent.

If a diode bridge is incorporated into the circuit, a direct response in absorbance units can be obtained, thus saving the labor of converting the arithmetic response into a logarithmic one.

A readout device can incorporate several other desirable features. One is signal integration, which is accomplished by collecting signal response over a longer period and then averaging the results electronically. This evens out instantaneous variations. Another is scale expansion. The readout signal can be recorded on a scale expanded as much as $100\times$, so that a 1% signal can be made to occupy full scale. This makes it possible to read the signal to much greater precision, although it must be remem- bered that noise is also expanded.

Other operating aids are automatic zero setting at the push of a button, automatic background correction, and direct presentation of concentra- tion. These accomplishments of the electronics engineer, using solid-state devices, reach their peak in digital readouts, which can incorporate all of the foregoing features and also contain circuits to linearize working curves that depart from straight lines. Curvature of the working curve occurs at higher concentrations and can introduce significant errors. How important curve rectification can be when the analyte is a major compo- nent of the sample is well illustrated in the paper by Fernandez and Kerber (182).

With a digital readout the results can be logged by an accessory sequencing printer or a programmable teletypewriter.

The engineers have not yet learned how to build an automatic chemist, but they have succeeded in removing most of the sweat from analytical chemistry—and, I fear, most of the fun.

3.3 EXCITATION SOURCES

The primary characteristics looked for in the excitation source are stability of output, sharp-line emission, and good intensity.

The importance of stability should be obvious. As the AAS method is serial—one determination at a time—the source output should not change between the time when the 100% and zero controls are set and the time when standards and the last unknown are run. Stability of output depends not only on the quality of source design but also on the quality of the power supply, which must be carefully regulated.

Temperature change is the chief cause of instability. Gas discharge lamps require time to reach temperature equilibrium with their surroundings; this time could be shortened by good design of the source enclosure. Another expedient is to put the source on continuous warmup. A double-beam illuminating system will compensate for changing emission, but this should be trusted only after test.

Emission intensities of discharge lamps can usually be changed by changing the lamp current, but an increase in current shortens the lamp life and causes broadening of the emitted spectrum lines.

Lamp construction should be sturdy, and the lamp, whatever its type, should be capable of long life. This affects cost, which can be an important factor; each element measured requires its own source, and the cost of an extensive library can be considerable. Within the tube, the light path from the source should not traverse a cloud of cool gas, or a portion of the radiation will be self-absorbed, resulting in a broadened line and reduced intensity. Finally, the design should be such that metallic vapors do not corrode the internal structure or deposit on the inside surface of the window. For work in the ultraviolet this window must be of fused silica.

The sources most used in AAS are hollow-cathode lamps and high-frequency electrodeless discharge tubes. For certain elements and certain special purposes, Geissler tubes, xenon lamps, and mercury vapor lamps are occasionally mentioned in the literature, but for general problems these have no advantage over hollow-cathode lamps or ·electrodeless discharge tubes.

3.3.1 HOLLOW-CATHODE LAMPS

The hollow-cathode lamp is a low-pressure gaseous discharge tube. It differs from other such discharge tubes as neon sign lights, Geissler tubes, and the ordinary fluorescent lamps in the design of its cathode, which has a cylindrical cup-shaped conformation (Fig. 3.7). This form causes a high ion concentration, impinging on the inner wall of the cup at high velocities with the ejection of atoms of the cathode material by a process not yet fully understood. This results in a glow discharge and the emission of a low-energy spectrum of the material.

The low pressure and low temperature (about 500 to 600°K) result in little Doppler broadening and therefore very narrow spectrum lines. The discharge can be made very stable by close control of the power supply. These two properties of sharp lines and stable emission make the lamp ideal for AAS purposes.

The presence of hydrogen gas is particularly detrimental to performance because it reduces the emission intensity; the intense emission of a hydrogen continuum in the ultraviolet dilutes the desired emission, reducing signal intensity and causing curvature in the working curve. The source of the hydrogen is the highly purified metal used for the cathode

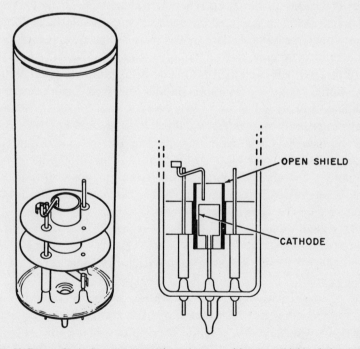

Fig. 3.7 Hollow-cathode lamp construction. A type with open shield of the cathode (Perkin-Elmer patent).

material; these purified metals are usually prepared electrolytically, a process that invariably occludes hydrogen. Thorough outgassing during manufacture is obligatory.

Another difficulty is caused by the low melting point of some metals. Some, like tin, are not very volatile although low in melting point, and are difficult to sputter into the cathode space in sufficient concentration to yield a satisfactorily intense emission. Using intermetallic compounds in such a case, with the added metal the more refractory, allows the use of higher lamp current and thus improves intensity.

Certain lamps are made to be operated at currents high enough to melt the cathode metal. The melt wets the cathode surface but does not flow out of the cup. Such lamps for the low-melting metals tin, gallium, indium, bismuth, and lead have up to ten times the brightness of conventionally constructed lamps with these metals. As each kind of lamp has its own peculiar characteristics, it is well to follow the manufacturer's instructions in its use.

Electrical requirements for hollow-cathode lamps are not severe. With either alternating or direct current, the voltage across the lamp during operation is 200 to 300 V. Striking voltage for ignition is about 400 V, and in order to have some ballast in the circuit the power supply voltage should be 500 or more volts. With alternating current the emission is pulsed at each half-cycle corresponding to cathode negative. With full-wave rectification, the emission follows the frequency of any ripple remaining unfiltered.

Current requirements differ with each kind of cathode. Operating and maximum currents are carefully specified by the manufacturer, and guarantees can be voided if these limits are exceeded. Maximum current does not exceed about 50 mA, so this is the minimum current that the supply should be capable of delivering. The operating currents for dc and ac use are not the same; with alternating current the root-mean-square current should be lower.

Lamp Envelope and Window Material

The window material for lamps whose resonance lines fall in the short-wavelength region must transmit in this region. Borosilicate glasses (Pyrex), of which the body of the lamp is customarily made, transmit satisfactorily down to about 280 nm, but below that only fused silica is practicable. Fused silica cannot be connected directly to borosilicate glasses because of the difference in temperature coefficients. Early lamp designs resorted to cemented window seals, which proved unsatisfactory because of the tendency to leak. Modern lamps are made with a graded seal between the window and the glass of the lamp body.

Transmission requirements for the principal resonance line of a particular element are not the only consideration. The weaker, secondary resonance lines, which are often very useful (for example, when dealing with high analyte concentrations) and may be at shorter wavelengths, must also be considered in choosing the window material.

Lamp bases are of molded plastic, similar to radio tubes, but it should be noted that they are not standardized. Lamps should be bought from the supplier of the AA instrument; if from a different supplier, then an adapter must be used.

Fill Gas

The physical properties of the fill gas (the ionization potential and the atomic weight), together with the pressure, the cathode geometry, and the applied current interact in a complicated way to affect operating characteristics, with differing effects for different metals and indeed for different spectrum lines of the same element. Mitchell (44) has studied this problem. Both the neutral and ionic spectra of the cathode metal are emitted, but only the former is of value and only the low-energy lines are best. Based on many empirical experiments, the fill gases most used are argon and neon, with the latter preferred in recent years.

Crosswhite, Dieke, and Legagneur (45) studied the interaction of fill gas, pressure, and operating current on emission stability to determine the best combination. They found that neon was preferable, that the optimum pressure was 3 mm Hg, and that argon, while more effective as the exciting medium, caused more self-absorption and therefore produced a lower intensity. Their cathode metal was iron.

L'vov (46), describing some Russian work that may be unfamiliar to Western workers, gives an excellent exposition of the influence of the fill gas. He states that for maximum intensity, the pressure should be in the range 2 to 5 torr, with change of internal pressure evidenced by a change in brightness. The kind of gas affects even the brightness of individual spectrum lines; this is because of the difference in ionization potential of the various fill gases and therefore in the electronic temperature of the discharge. The relationship connecting intensity with current, atomic species in the gas, and cathode element can be expressed by the equation

$$\text{intensity} = ai^n$$

where i is the current and a and n are constants of the cathode element and gas, respectively.

Life of Hollow-Cathode Lamps

The life of a hollow-cathode lamp is related to the maximum current used in its operation, but its intensity is related to the average current. If a dc supply is used (with mechanical chopping), the two are equal. With an ac supply (pulsed half-wave), peak current is more than twice the average, thus reducing life unnecessarily.

Failure of the lamp occurs when the fill-gas pressure has been reduced by cleanup to the point where the lamp can no longer be started and must be discarded. The cleanup process is due to adsorption of the gas on metallic particles sputtered from the cathode (not to impurities in the cathode metal, as some workers believed). An exhausted lamp may sometimes be started a few more times by priming with a Tesla coil.

The volume of the lamp envelope has some effect on lifetime. Up to a point, the larger this volume the longer the lifetime, because more gas is available (47).

A 1964 report by Westinghouse (48) on lifetimes of lamps of their manufacture indicated that these lifetimes varied from 28 to 65 A hr. The lamps were neon-filled, cathodes were made of common metals, and currents varied.

A good indication of (at least) minimum lifetime is the guarantee issued by manufacturers to cover their products. A typical one reads "Lamps are guaranteed to emit the spectra of the elements indicated for 5000 milliampere-hours or for two years from the date of shipment, whichever occurs earlier. The manufacturer further guarantees that during the first six months after shipment or 5000 mA-hr, whichever occurs earlier, the lamps will meet or exceed the intensity and absorption specifications to which they are tested. This warranty is voided for lamps which sustain physical damage or are run above rated currents." We may conclude from all this that lifetimes can be expected to greatly exceed the 5000 mA hr of the guarantee.

Multiple-Element Lamps

Of great interest to the analytical chemist is the development and availability of lamps fitted with cathodes containing more than one element. These reduce the cost of performing atomic absorption analyses and the cost of the lamp library. Many workers have attempted to make hollow-cathode lamps that work equally well for more than one metal. Usually, lamps made of metal alloys initially emit the spectrum of each constituent, but gradually the more volatile metals sputter from the surface, leaving the cathode rich in the remainder. Thus, alloys are not favored for multiple-element lamps.

At first, only partial success (49) was attained in building these multi-element lamps, using various expedients. An effective technique has been to fabricate cathodes of mixed metal powders. The powders are pressed in a die and sintered at an appropriate temperature. By this means lamps have been made that work equally well for such disparate metals as copper and manganese, and copper and chromium. A lamp emitting the spectra of six elements has been described (50). Fine powders of high-purity metal were mixed together in the proportion of 25% copper and 15% each of manganese, chromium, iron, nickel, and cobalt. The mixture was pressed in a die designed to yield a solid cylinder. The compacts so formed were heated to a temperature somewhat below the melting point of copper. The cathode was then machined and assembled into the lamp by conventional procedures, and the lamp was filled with neon. Figure 3.8 shows that performance was not very different from single-element lamps.

Another approach to multielement lamps utilizes true intermetallic compounds (51); this technique also proved practicable.

Multielement lamps are not without problems. If two elements have resonance lines that are not resolved by the monochromator being used, an apparent interference results. Some combinations of elements appear to be incompatible. Emission intensities of the elements relative to each other gradually change with age. Two or more elements in the same envelope impose compromises in choice of fill gas, window material, and operating characteristics.

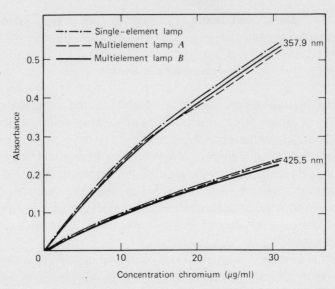

Fig. 3.8 Comparison of performance between a single-element chromium lamp and two multielement lamps containing chromium. Aside from requiring an increased current for the latter, the differences are immaterial.

Desirable characteristics of multielement lamps are easily listed. They cost less than the sum of the equivalent single-element lamps. The time saved in not requiring a lamp change for each element change could be significant in the case of single-channel instrumentation with no warmup accessory.

Multielement hollow-cathode lamps are now a standard commercial item. They fit very well to the analytical problems of many large industries; for example, some of the industrially important combinations are chromium, cobalt, copper, managanese and nickel for the steel industry; a combination of calcium and magnesium for the cement industry; a combination of chromium, cobalt, copper, iron, manganese and nickel for the glass industry.

Demountable Lamps

In a laboratory where many elements are being determined, the library of lamps that must be stocked can be a heavy expense. The temptation then is to use demountable lamps, in which only the cathode need be changed when changing from one element to another. The cleanup of fill gas and therefore lamp life does not apply, as the lamps can be purged continuously or temporarily sealed. But this scheme has certain disadvantages. In comparison with sealed commercial lamps, demountable lamps require a much longer setup time and are usually not as bright. They also require skill and experience in vacuum technique for their making, and the availability of vacuum equipment.

3.3.2 ELECTRODELESS DISCHARGE LAMPS

Second in popularity to hollow-cathode lamps for AAS are electrodeless (high-frequency) discharge tubes. The driving energy for these EDLs in the past has been mostly the Raytheon medical diathermy unit, probably because it was available as an operating system with no need for modification. Its output was about 125 W at 2450 MHz. However, the supply to the unit was unregulated and stability was so poor as to make it unfit for quantitative work. Recently, Brandenberger (52) reported on how to make a minor modification in the power supply circuit to produce a stable output.

Other power supplies, down to about 10 MHz (the radio-frequency range), were also reported in use. Indeed, short-wave radio transmitters of 100 to 150 W output could also generate the necessary high frequency.

Commercial units designed specifically for AAS are now on the market. Both lamps and power supplies are guaranteed to equal or exceed the stability of hollow-cathode lamps.

EDLs must be used in a resonant cavity or with a tuned antenna to transfer the microwave or radiofrequency radiation into the discharge tube. The chemical elements that work well in an electrodeless lamp must have sufficient vapor pressure at the operating temperature (about 500°C) to sustain the discharge. Elements of insufficient volatility in their free state can be used in the form of their halide salts, preferably the iodide. Fill gas to start the discharge and heat the tube can be helium, neon, or argon, but when temperature equilibrium is established the gas emission is at a low level and causes very little interference. Operating pressure is about 1 mm Hg. At this pressure and temperature, the spectral line emission is sharp, with intensities much better than the corresponding hollow-cathode lamps.

Fortunately, the elements that cause trouble in manufacture or use in hollow-cathode lamps are just the ones that work best as EDLs. For example, Perkin-Elmer supplies lamps for such troublesome elements as phosphorus, selenium, tellurium, aresenic, lead, cadmium, and tin (among others) in completely packaged assemblies (see Fig. 3.9 for internal construction) that are easily fitted into the standard holders for their hollow-cathode lamps and are then positioned correctly on the optical axis of the instrument.

Power supplies for hollow-cathode and electrodeless discharge lamps are not interchangeable, an inconvenience when both must be used for the same sample. Electrodeless lamps are somewhat more expensive than the corresponding hollow-cathode lamps but have a longer life.

Details on the operation of EDLs are given in references 53 and 54. L'vov more recently compared the EDL and the hollow-cathode for AAS work (55).

3.3.3 OTHER EXCITATION SOURCES

Vapor discharge lamps, of the type exemplified by the German Osram lamps, have had a limited use for the excitation of the alkali metals and of some of the other high-volatility elements. These lamps operate at much higher pressure than either hollow-cathode lamps or EDLs; consequently their line emissions are much broader, although their intensities are high.

Simple discharge tubes such as the low-pressure mercury penlamp are practicable for AAS. Because of the low pressure, lines are narrow, intensity is adequate, and power supplies are simple. The small mercury lamp is still the common source for the cold atomic absorption method for determining Hg. Some of the other low-melting elements, such as gallium and cadmium, could be used. The tubes are easily made in the laboratory.

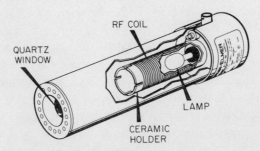

Fig. 3.9 High-frequency electrodeless discharge lamp showing internal construction. An RF tuned coil surrounds the emitting chamber containing noble gas and a compound of the element whose spectrum is sought. (Perkin-Elmer patent)

Slavin (56) found that when it is operated on a well-regulated current supply, either ac or dc, the mercury penlamp becomes constant in output when the tube temperature reaches equilibrium with the ambient air. No doubt lamps of the other elements will do likewise.

An interesting possibility as an excitation source is the tunable laser beam, which is still in the development stage, although Neumann and Kriese (57) have reported on its use for atomic fluorescence. They reported that they were able to improve sensitivity by an order of magnitude.

The laser beam contains only a few, well-separated lines, so spectral interference is no problem. Over its tunable range it can replace the individual element lamps within that range, with a saving in lamp costs.

The laser's extremely narrow beam, with practically no divergence, should be very useful in illumination systems. For example, the conventional system using the cathode of a hollow-cathode lamp as the source, and either lenses or mirrors to direct the beam through the graphite tube of a furnace, raises insoluble problems in geometric optics. The aperture of an absorption cell, with an aspect of 4 or 5 times the diameter, is such that a conventional system is extremely inefficient. With a laser, tube furnace illumination can be very efficient, and long paths from flame atomization would also be possible. The high intensities that can be concentrated in a very narrow line should increase sensitivity markedly.

The status of research in laser technology at the present writing is such that this may be the new excitation source, replacing the hollow-cathode lamp. This depends entirely on the development of the tunable dye laser, whose emission is in the form of widely spaced, extremely sharp lines, continuously changeable in wavelength, hence the term "tunable." A tunable laser would simplify equipment and increase sensitivity of measurement.

3.4   FLAMES

A flame burning in air forms an ideal means for converting a solution into the atomic vapor required for atomic absorption. While a flame presents certain problems and disadvantages, the advantages are so compelling that it is unlikely that any technique will completely replace it. A flame is simple, inexpensive, and very easy to use. A flame provides a remarkably stable environment for atomic absorption. The fact that determinations can be obtained to 0.2% relative precision means that the atomic population in the light beam can be reproduced to a few parts per thousand.

The successful utilization of flame technology has been largely empirical. Although their detailed mechanisms are still not well understood, flames are easy to use, and we do have a general understanding of their physics and mechanics as they are used for atomic absorption spectroscopy. We are thus able to predict what will work well for a particular problem. It also is possible to delineate the remaining problems of flame control and to outline the potentially useful avenues for research.

The function of the flame is (1) to evaporate the solvent, (2) to decompose and dissociate the molecules, and (3) to provide ground state atoms for absorption of the exciting radiation. Of these functions, the last, if it were possible, should bring all analyte atoms to the absorbing condition and no more. This is impossible because of two opposing conditions, which force a compromise. The flame temperature must not be too high, otherwise a fraction of the atoms will be ionized and so lost to absorption. But it must be high enough to break the molecular bonds. Furthermore, the kinds of flame gases should be kept to a minimum, no more than two or three, for the sake of simplicity.

The fuels commonly used in flames are propane or ordinary coal gas, hydrogen, acetylene, and, infrequently, butane. Oxidants are oxygen or oxygen with diluting inert gas, air, and nitrous oxide.

A mechanical requirement is a balance between the velocity of propagation of the flame and the velocity of gas emerging from the burner tip. If gas velocity is too high, the flame will lift off and burn above the tip; if too low, there is danger of flashback. A desirable property of the mixture is the ability to change the fuel–oxidant proportions to obtain either an oxidizing or a reducing flame and still encounter no troubles of flashback or liftoff.

Certain metallic oxides are so tightly bound—for example, A1O, ZrO, UO, TaO, and NbO—that they require a high-temperature flame, but pure oxygen cannot be used safely except in a burner unsuitable for AAS. However, there are gas mixtures available that can be burned safely in AAS burners. Some gas mixtures that have been suggested in the literature have been discarded because they are poisonous, are hard to obtain, or require premixing.

Table 3.1 lists the characteristics of some flame mixtures with respect to their velocity of burning and their temperatures. Comments on the more generally used mixtures follow.

TABLE 3.1. Characteristics of Some Flame Mixtures (60)

	Flame Velocity (cm/sec)	Temperature (°C)
Air–propane	82	1925
Air–acetylene	160	2300
50% O_2–50% N_2	640	2815
Oxygen–acetylene	1130	3060
Nitrous oxide–acetylene	180	2955
Nitric oxide–acetylene	90	3080
Nitrogen dioxide–acetylene	160	—

Air–Acetylene

This is the general-purpose fuel for AAS work; it is effective for the atomization of about 35 elements (58, 59) and presents no danger of flashback. Temperature of the flame is about 2300°C, somewhere in the middle range. It is characterized by low electronic noise and low interference effects. The flame mixture can be varied from oxidizing to reducing.

However, for those elements whose oxides are difficult to dissociate, air–acetylene cannot produce the necessary sensitivity to be of practical use. For this, nitrous oxide–acetylene, with its higher temperature of burning, must be used.

Nitrous Oxide–Acetylene

Willis (60), working on the problem of atomizing the refractory elements, reported in 1965 that the nitrous oxide–acetylene flame had a sufficiently high temperature (about 2900°C) to atomize those refractory elements that were beyond the scope of air–acetylene. While the temperature approaches that of oxygen as the oxidant, the flame velocity is so low that flashback dangers are almost absent. In tests of some of the refractory elements, he reported sensitivities of 0.03 to 5.0 μg/ml. This was an outstanding contribution to the developing technique of AAS and opened a large segment of the periodic table to successful analysis. Happily, while nitrous oxide could not often be found in an analytical laboratory, it was obtainable from dental supply houses in pure form, and so was readily available to experimenters. Experience with N_2O over the years since Willis' publication has shown that the increase in sensitivity for such difficultly determinable elements as aluminum, barium, beryllium, silicon, and many others has been marked. Allen lists (59) the free-atom fraction of total atoms in the nitrous oxide–acetylene flame compared to that existing in the air–acetylene flame. This is shown for 27 common elements in Table 3.2.

Air–Hydrogen

An air–hydrogen flame burns at the low temperature of about 2000°C. It is almost invisible, although the traces of sodium compounds invariably

TABLE 3.2. Free-Atom Fractions in Air–Acetylene and Nitrous Oxide–Acetylene Flames (59)

Element	Air–Acetylene	Nitrous Oxide–Acetylene
Ag	0.70	0.57
Al	0.00005	0.13
Au	0.40	0.27
Ba	0.0018	0.17
Be	0.00005	0.095
Bi	0.17	0.35
Ca	0.07	0.52
Cd	0.38	0.56
Co	0.28	0.25
Cr	0.071	0.63
Cu	0.88	0.66
Fe	0.84	0.83
Ga	0.16	0.73
In	0.67	0.93
K	0.32	0.12
Li	0.12	0.34
Mg	1.0	0.88
Mn	0.62	0.77
Na	1.0	0.97
Pb	0.77	0.84
Si	0.001	0.06
Sn	0.043	0.82
Sr	0.063	0.26
Ti	0.001	0.11
Tl	0.52	0.55
V	0.15	0.32
Zn	1.0	0.60

present in solutions and in room air reveal the flame's presence. Because of the low temperature, this flame is especially suitable for the alkali metals, although it is not used much for this purpose. The presence of hydrogen in combustible mixtures always presents the danger of flashback and explosion. The air–hydrogen flames should be used in burners designed for it, and it must be both ignited and extinguished according to the instructions of the burner manufacturer.

Argon–Hydrogen

The particular virtue of this flame is that with air as the oxidant it exhibits very low absorption in the shorter ultraviolet (below about 220 nm) and so is useful for the determination of arsenic (wavelength 194 nm) and selenium (wavelength 196 nm). The flame temperature is very low, about 300 to 800°C, depending on the portion of the flame used, so that chemical and matrix interferences are more likely than with the higher temperature flames. These interferences can be reduced or eliminated by using the method of additions or by matching closely the compositions of standards and unknowns.

As with air–hydrogen, the argon–hydrogen flame is invisible but will be colored yellow if a trace of sodium is present in the solutions. This gas mixture requires a specially designed burner head.

Figure 3.10 shows the absorption of the flame alone in the short-ultraviolet, together with the air–acetylene and air–hydrogen flames; it can be seen that the argon–hydrogen flame is superior to acetylene and much better than hydrogen alone below 2100 Å.

Air–Propane (Natural Gas)

This fuel has been found by some workers to be useful for certain elements that require a low temperature. Its flame temperature is about 1925°C, but interferences are more likely because of this low temperature.

Flames for chemical analytical purposes have been studied extensively and a large literature exists. Willis (61) has published a review article; de Galen and Samaey (62) and Chester et al. (63) have discussed flame gases. Dean (64) has a good chapter on the general subject in the text edited by him and Rains. Suddendorf and Denton (65) discuss burning parameters of flames of oxygen with acetylene and hydrogen. Reynolds and Lagden

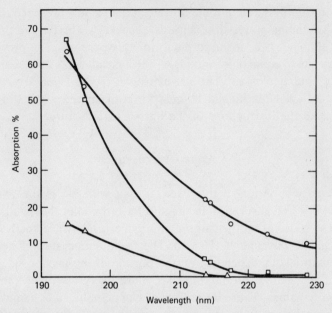

Fig. 3.10   Absorption in the far-ultraviolet region by flames of (○) air–acetylene, (●) air–hydrogen, and (x) argon–hydrogen. At 200 nm the argon–hydrogen flame absorbs at only 10%, compared to 30% for air–hydrogen and 45% for air–acetylene.

(66) point out certain possible hazards when using acetylene with silver salts.

3.5   BURNERS AND NEBULIZERS

The first burners used in AAS work were direct-injection burners, as exemplified by the Beckman burner (Fig. 3.11). This consisted of a vertical cylinder with two concentric chambers into which the fuel and oxidant were passed. The two gases came in contact at the burner head, and the shape of the flame was columnar, ideal for illuminating the slit when used for emission, but providing only a short path for absorption. Because the gases did not mix until they left the device, the direct-injection burner could be used with oxygen as the oxidant, with fair assurance against explosion; oxygen provided the high flame temperature necessary to dissociate refractory compounds.

The direct-injection burner has several faults when used for AAS. The

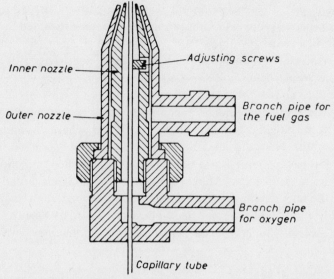

Fig. 3.11 The Beckman burner, representative of the direct–injection types.

flame is turbulent and noisy, both electronically and acoustically. The sample is injected directly, providing no opportunity to remove the larger droplets of liquid; consequently they pass through the flame unatomized and indeed still solvated. Dean and Carnes (67), Gibson et al. (68), and Winefordner et al. (69) all showed this to take place. Efficiency of atomization is estimated to be only about 20%. A review of the subject can be found in Burgess (70).

Willis' paper (60) on nitrous oxide appeared in 1965 and was soon followed by a report by Amos and Willis (71), showing that this oxidant, with such a fuel as acetylene, could be mixed within a burner with very little danger of flashback. The flame was about 600° hotter than air–acetylene and about the same as for oxygen–acetylene.

These properties of high flame temperature and the possibility of premixing allowed for the design of a burner that was much more suitable for AAS—the premix burner, which has now largely replaced the direct-injection type.

The premix burner was used in Australia and New Zealand on the early development work for the AAS method. The gases were mixed in a chamber before reaching the burner head, and the flame was shaped in a long, thin rectangle aligned along the optic axis. This geometry provided a long optical path for absorption, similar to the path in the cell used for molecular spectroscopy. A commercial nebulizer was included in the gas train. Various workers since then have modified the basic design to increase resistance to corrosion (72), to increase sensitivity (73), and to reduce electronic flame noise (74).

Figure 3.12 shows the construction of a modern burner-nebulizer assembly. The sample solution is taken up by a slight vacuum into a fine capillary tube; it then enters the nebulizer chamber, where it is mixed with the flame gases. Here the larger droplets settle out and flow to waste. The aerosol and gases then pass to a spoiler or baffle, whose function is to remove the finer droplets. About 85% of the sample is discarded in this way. By the time they reach the burner, the aerosol, fuel, and oxidant are thoroughly mixed. The burner head consists of a solid block of either stainless steel or titanium metal (which is easier to machine) in which a

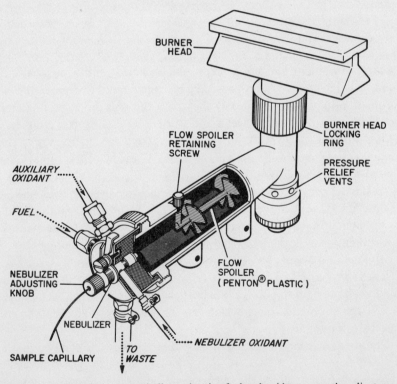

Fig. 3.12　A commercial burner-nebulizer, showing fuel and oxidant ports, the adjustment for sample uptake, nebulizer chamber with flow spoiler, and a single-slotted burner head.

long, narrow slot has been cut. The width of this slot is designed to balance the flame propagation velocity against flow velocity of the gas, as too wide a slot increases the danger of flashback and explosion of the mixed gases. The slot width, therefore, should be chosen to fit the gas mixture. For air-acetylene, for example, a width of about 0.025 in. appears to be safe.

A three-slot burner was suggested by Bolling (75). This could provide for narrower slots and therefore be less subject to flashback, while permitting the same or greater throughput of sample solution than a single-slot burner. Thus, there would be less tendency for solutions with a high salt concentration to clog the burner. The flame is wider and its geometry can be more easily matched to the optical system employed. This is a point not often mentioned, but Kirkbright and Sargent (76) state it in this way: "The optimum flame geometry is merely that which places as many sample atoms as possible in the path of the light beam per unit time as it travels from the source to the detector." Multiple-slot burners are available commercially.

Premix burners will handle solutions of dissolved solids up to several percent without clogging. Burner heads consisting of a row of fine holes, proposed by some workers, are in general not favored because of the difficulty of cleaning them. The flame in a slot burner, owing to the smooth laminar flow of the mixed gases through the slot, burns very quietly and evenly, with very little noise, either audible or electronic. A description of design and performance is given by Willis (77).

Accessory equipment for the burner system has become sophisticated. Besides sensitive flowmeters to control and indicate the flow rate and pressure of gases, valves are electrically activated, providing the opportunity to switch from one gas mixture to another by pushbutton. The order of flow of the components in the mixture and interlocking with the proper burner head can be arranged by the electrical control.

Burner and nebulizer chambers have been carefully designed to be corrosion-proof against the mineral acids contained in sample solutions and are easily disassembled for cleaning.

3.6 THE FURNACE OR FLAMELESS ATOMIZER

A high-temperature resistance furnace of graphite or carbon, protected from oxidation by being bathed in an inert gas, can be used as well as the flame for dissolution and atomization of samples. The idea is by no means new; King described just such a furnace in 1908 (78). As carbon has no liquid phase, but sublimes at about 3800°K, temperatures in a carbon furnace can exceed flame temperatures.

This use of the high-temperature furnace was first suggested by the Russian scientist B. V. L'vov in 1959. He later elaborated on its application for AAS, in a long section of his monograph on atomic absorption (79), stressing the difference in measuring absorption in a graphite tube or cuvet compared to the flame. A diagram showing the arrangement of the cuvet and sample holder is shown in Fig. 3.13. The tube (2) is held at its ends in large graphite blocks (3) that have been drilled to provide holes concentric with the tube. These end blocks make the connection to the power supply, which is a 4 KW transformer with a secondary voltage of approximately 10 V.

The sample is placed in a cavity bored into the end of a tapered, necked graphite rod (1) such as is used in an emission arc. The tapered end fits into a tapered radial hole in the cuvet and so introduces the atomized sample into the exciting beam. The sample holder is heated by a separate, smaller transformer. The whole assembly is contained in a gastight hous-

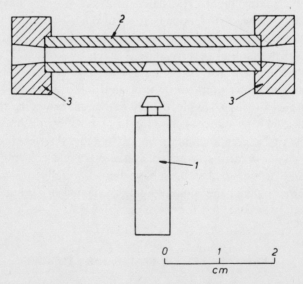

Fig. 3.13 The L'vov atomizer furnace. The cuvet (2) is about 4 cm long, with an internal diameter of 2.5 mm. The electrode containing the sample in a small cavity in the end is shown at 1, and its manner of fitting to the cuvet is clearly indicated.

ing with ports for the exciting beam and connections for inert gas and cooling water. Pressure within the housing can vary from atmospheric to several atmospheres. The apparatus can thus serve to trap noxious fumes or radioactive gases.

Workers objected to the need for two sources of heat, so several modifications of the L'vov design were soon published. The most popular of these is the arrangement of Massmann (80) shown in Fig. 3.14. It uses the same resistance-heated tube, but the sample is introduced by means of a micropipet through a small axial hole in the tube wall. It should be noted, however, that L'vov criticizes Massmann's simplification on the grounds that it increases fractional vaporization of the sample and gives increased background and lower throughput because liquid samples must be dried in the furnace, whereas separate electrodes can be dried in batches on a hotplate.

To overcome this last objection to the Massmann furnace, Hwang et al. (81) proposed a rectangular cuvet with a slot cut in its side, into which a microboat holding the sample is inserted. The microboat, $10 \times 8 \times 1$ mm and holding a maximum of about 80 μl of sample, could be dried or charred outside the furnace. Hwang et al. pointed out that an additional advantage of the microboat is the possibility of treating certain intractable

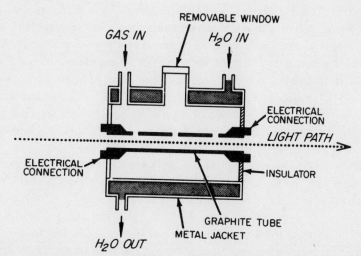

Fig. 3.14 Cross section of furnace embodying Massmann's idea. Inert gas surrounds the heated tube. Water-cooling of the furnace jacket is provided, to cool the cuvet rapidly after atomization. The sample solution is introduced into the cuvet through a hole at the center, after removing the inspection window.

samples. They illustrated this by describing procedures for analysis of vanadium in crude oils, iron in pitch, and titanium in polypropylene. In general, samples with matrices that are difficult or impossible to weigh directly can be weighed by difference, using the microboat.

Furnace atomizers are now offered by manufacturers as an optional accessory compatible with their instruments. Cuvets are about 30 mm long, with inside diameter about 8 mm and a sample hole of about 2 mm diameter. Maximum volume that can be used is about 100 μl. Solid samples can be introduced by exposing one end of the cuvet and inserting a spoon carrying the weighed sample into the cuvet. About 0.1 to 0.2 mg of solid sample can be accommodated. Cuvets are now coated with a layer of pyrolized graphite to minimize the absorption of liquid and keep it on the surface.

The required capacity of the power supply is 4 to 5 KW, and with this the maximum temperature attainable is about 3000°C. This temperature, it will be observed, is approximately at the level of the nitrous oxide–acetylene flame and considerably higher than the temperature of the air–acetylene flame. To protect it from oxidation when hot, the graphite surface is bathed continuously in inert gas, usually argon or nitrogen. The cuvet is water cooled during addition of sample.

A smaller version of the Massmann furnace (the mini-Massmann) has been described by Matousek (82), and others have also endeavored to miniaturize the parts in order to cut down the power requirements to the point where connection could be made directly to ordinary 15 A wall outlets. Proposals to improve the cuvet or other means of holding the sample include using a solid graphite rod in place of a tube (83), graphite braid (84), or ribbons of refractory metal (85) such as tantalum or tungsten. Siemer et al. (86) describe a simple technique for coating the graphite surfaces of the atomizer chamber with pyrolitic graphite to prevent diffusion of analyte vapor through the cuvet wall.

The absorption signal in furnace atomization, obviously, is not constant with time like the flame signal, but is changing very rapidly and must be measured at a peak. This requires a recorder or other measuring device with a very short period, at most 0.1 or 0.2 sec. In addition, an indicating meter must have some provision for holding the peak reading.

Recent reviews on furnace construction and operation have been published by Woodriff (87) and by Koirtyohann and Wallace (88). A good general treatment of flameless atomization can be found in the chapter by Massmann (89) in the textbook edited by Dean and Rains.

3.6.1 TEMPERATURE PROGRAMMING

Soon after the furnace technique of atomization was developed, it became evident that some sort of temperature programming was needed if reproducible results were to be obtained. The programmed stages would be four—drying, charring, atomizing, and a final stage of heating at high temperature to remove from the cuvet all traces of sample in order to prevent memory effects. As the task of programming proved onerous, the obliging makers of instruments automated the whole procedure, with control by the inevitable pushbutton.

The heating program is depicted in the diagram of Fig. 3.15. The rate of rise of the cuvet temperature, together with residence time at each temperature, can be set and reproduced by controls. With the automatic feature, reproducibility can be maintained and settings repeated to match requirements of samples according to their different characteristics, both physical and chemical.

Direction of the gas flow through the cuvet is also of importance. Figure 3.16 shows one manufacturer's design; the purpose is to increase retention time of the analyte vapor in the optical path and also to reduce background. The cause of the latter is the deposit of fume on the cooler ends of the cuvet during the ashing process; the deposited material is then atomized again when the heat is increased, adding to the background signal. As shown in Fig. 3.16, the gas flow carries the vapor toward the

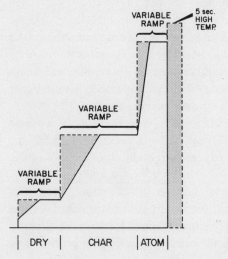

Fig. 3.15 Diagram of temperature or ramp programming of the furnace. The four stages of heating are shown as steps.

CROSS SECTION

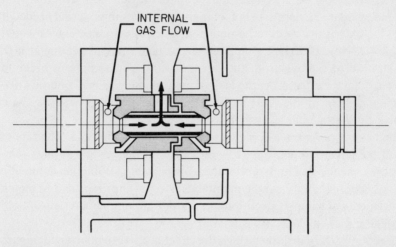

INTERNAL
GAS FLOW

Fig. 3.16 Cross section of a commercial furnace. The inert gas is introduced at both ends of the cuvet and flows out at the sample hole, filling the chamber within the outer case.

middle of the tube and out at the sample hole. An aid in increasing sensitivity is a stop-flow device that enables the operator to stop the flow of purge gas momentarily during the signal-measuring interval.

3.6.2 BACKGROUND CORRECTION

Background correction is needed because the exciting radiation in the fume is scattered by some molecular species, by salt particles, and by smoke particles. Without correction, this increased absorption will cause high results, and it cannot be eliminated by conventional double-beam arrangement of the optics or by an ac excitation source.

An accessory to make the correction automatically is a device that employs a deuterium or hydrogen lamp with its continuum radiation. Correction is not strictly effective under all conditions, but it is quite satisfactory on the whole.

The accessory works in the following manner. A continuum beam from the deuterium lamp is inserted into the optical path by, for example, the chopper, which produces a rapid alternation of the deuterium and exciting beams at a frequency to which the amplifier is tuned. After both beams pass through the vapor in the cuvet (or the flame) and reach the detector,

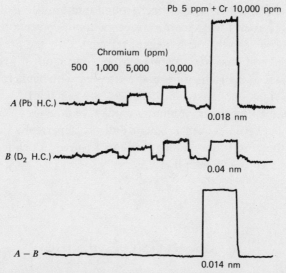

Fig. 3.17 Deuterium lamp background correction. Recorder trace *A* is of the sample absorption alone. Trace *B* is of the deuterium lamp alone. Trace *C* is the result of electronically subtracting trace *B* from trace *A*. (Courtesy of Instrumentation Lab., Inc.)

the deuterium signal can be subtracted electronically from the exciting beam signal, which contains the sum of background and atomic absorption signals. A recorder trace (Fig. 3.17) with and without correction will make the operation clear.

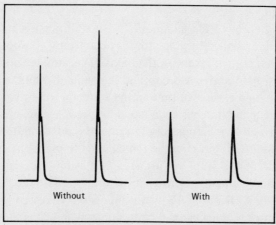

Fig. 3.18 Correction for fume interference. (Left) Atomization of the sample, copper in seawater, causes fume to appear in the cuvet because of the high salt concentration, thus obscuring the copper peaks. (Right) With background correction, the fume spikes are removed.

Correction of another type of background interference is illustrated in Fig. 3.18. This shows two recorder traces without correction on the left. The spikes obscuring the peak heights are caused by the presence in the sample (copper in seawater) of a high concentration of salt, which is atomized at about the same time as the copper. With background correction, on the right, the spikes in the traces are eliminated.

Background by means of a continuum can, of course, be corrected by making the background measurement entirely separately and manually subtracting the absorption reading, but this assumes no change in conditions for the two readings and is unsafe.

3.7   THE DELVES CUP ATOMIZER

A cross between flame and furnace atomization is the sampling boat technique. The device, described in an early form by Kahn et al. (90) and later by Kahn and Sebestyen (91), consisted of a tantalum ignition boat that held the sample and was introduced into a flame, slowly at first to evaporate the solvent and then into the full flame. The boat's position was just below the optical axis. The signal was transient, meaning that the analyte's evolution was concentrated into a very short time, not at a steady state, as in conventional flame technique, and was therefore efficient of sample.

Delves (92) improved the technique by substituting a small nickel cup for the boat, and changed the heating procedure by positioning the cup just below a nickel tube on the optical axis. The sample vapor enters the tube through a hole above the cup and is scanned. Thus, the sample is used efficiently and residence time in the scanning beam is lengthened.

The Delves cup method has been much favored for the determination of lead in blood and urine, possibly because blood cannot be aspirated without great dilution, but now the tendency is to make these determinations by means of the graphite furnace. The Delves cup is restricted to easily volatilized elements. Sensitivity is quite good—equal to or better than conventional flame AAS—and precision is about 5%. This is sufficient for most biological analytical problems, and it is in this field that the Delves cup technique has had its greatest use. An advantage of the cup is that many samples can be dried and charred at the same time on a hotplate, so that throughput is good.

3.8 FLAME EMISSION METHODS

The two methods to be discussed in this section are direct atomic emission and atomic resonance fluorescence, both by means of flames. They really have no place in this book, whose title clearly restricts its field to absorption. However, atomic emission and fluorescence papers are always to be found in the same journals and often in the same textbooks as AAS, so perhaps it is appropriate to mention them here. In addition, instrumentation for all three methods is the same, except for some changes in the external illumination systems and some rearrangement in the detector electronics. A few journal references will be included, so that interested readers can pursue the subject.

My reasons for this limited treatment of flame emission are the following. Conventional AAS can do anything that flame emission can, and do it with greater precision, greater sensitivity, and much greater versatility. The successful method in emission spectroscopy, old and well-developed, uses the carbon arc and photography, or the spark and the direct reader, with the added advantage of multielement capability.

Resonance fluorescence, on the other hand, has arrived before its time. No important manufacturer offers fluorescence instruments as such. The field is still in the hands of academic researchers who design and build their own equipment; very little fluorescence work is being done on a routine basis. From present appearances, fluorescence has a promising future and may very well displace AAS for the group of elements for which it is especially suited.

3.8.1 FLAME EMISSION SPECTROSCOPY

Modern work started with a series of publications by Lundegardh (93) in the 1930s. He applied emission mainly to the analysis of rocks and minerals. For this application the flame was in direct competition with carbon arc spectroscopy, which at that time was in the midst of a vigorous development. Lundegardh's publications made very little impression on the spectroscopists of that period, for they could see no advantage in a procedure that required the laborious step of putting refractory compounds in solution when the sample for the arc could be used directly, only requiring grinding to a powder. Furthermore, the number of elements that could be induced to emit a spectrum was limited because of the low flame temperature produced by air–acetylene or air–illuminating gas, while the carbon arc could excite all of the metals and several of the nonmetals.

For the present time the status of flame emission has been reviewed (sympathetically) by Pickett and Koirtyohann (94) under the title "Emission Flame Photometry—A New Look at an Old Method." They point out that with the advent of hotter flames, better burners, and monochromators using gratings in place of prisms, the method has much greater versatility than before. They present a table of detection limits, which indeed shows improvement over Lundegardh's results. Too, a library of hollow-cathode lamps is not needed, eliminating a significant expense.

But all is not rosy. Certain inherent weaknesses of the flame emission method cannot be removed. Interelement interference due to mismatch between standards and unknowns is much more severe than in absorption. Difficulties in signal measurement are also greater; at low concentrations, electronic and background noise represents a large proportion of the total signal, so precision is poor. At high concentrations, self-absorption within the source causes a falloff of linearity of the working curve, so working range is short. Owing to the low temperature of the source, even with the new gases, spectral excitation is restricted to wavelengths longer than about 260 nm, thus excluding a valuable portion of the UV spectrum.

In addition, flame emission spectroscopy puts more severe demands on the monochromator than is the case with absorption methods. Extraneous light from the flame cannot be eliminated by the expedient of chopping and tuning the amplifier, as in absorption. In the latter, a single narrow line in a dark ground constitutes the recorded signal, whereas in flame emission all of the background light from the flame—molecular bands and continuum—within the bandpass of the monochromator goes on to the detector. Pickett and Koirtyohann suggest scanning on either side of the analyte line and then subtracting this radiation after averaging the level. Not only is this laborious, but the process is inexact, because the background trace is not smooth but consists of jagged peaks and hollows, making averaging little better than a guess.

Flame emission is especially sensitive for the alkali metals and for calcium, strontium, and barium, for which its greatest use is found. It is less sensitive although practicable for aluminum, gallium, and indium, for most of the rare earths, and for several of the transition metals.

Extended treatments of the flame emission method can be found in references 95 to 97.

3.8.2 ATOMIC FLUORESCENCE SPECTROSCOPY

This also is an emission method. Atoms illuminated by appropriate radiation can absorb the radiation and thereby be raised to an excited energy state. In returning to the ground state or to another level close to the ground state, the atom must release the energy it acquired during excitation. If this energy is released as radiation, we speak of the process as atomic fluorescence or resonance radiation. The intensity of the radiation is a direct measure of the analyte concentration in the flame and can be the basis for an analytical method.

Fluorescence radiation, like the collisional radiation in flame emission, is emitted in all directions equally. The exciting radiation is of no interest and can be disregarded. The optical arrangement is diagrammed in Fig. 3.19. The burner is mounted on the optical axis, but the exciting lamp is set at right angles so that none of its light will enter the monochromator slit. To ensure that light from the flame is distinguished from the fluorescence, the exciting beam is chopped and the amplifier turned to this frequency, as in AAS. Although the fluorescence radiation is diffuse, it can be collected in a beam of sufficient intensity to provide an adequate signal.

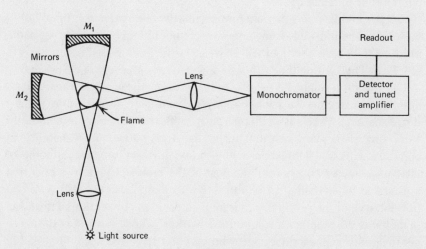

Fig. 3.19 The optical arrangement diagrammed for atomic fluorescence spectroscopy. The flame is on the optical axis, but the exciting lamp is off to one side, usually at right angles. Mirrors M_1 and M_2 serve to reflect back the radiation to the flame and to the monochromator.

As sensitivity depends directly on the intensity of irradiation of the flame, the high-frequency electrodeless discharge lamp has been found to be best. The lamp should be as close to the flame as safety permits, and the flame as close to the slit as possible, or its light collected and condensed on the slit by a lens or mirror. Cool flames, such as air–hydrogen, are preferable. Sensitivity is as good or better than AAS for many elements, and the working curve has a much longer dynamic range, often to 1000 times. For many problems, it is not necessary to disperse the beam, thus eliminating the monochromator. Interference filters may be used instead, with a large improvement in signal strength. Such an instrumental system is simple, rugged, and cheap, although some versatility is lost. Furthermore, electrodeless lamps have longer lives than hollow-cathode lamps.

The phenomenon of fluorescence was studied by R. W. Wood in the early 1900s, but it was not applied to practical analysis until 1964, when Winefordner and Vickers (98) published their pioneer paper. In the same journal issue, Winefordner and Staab (99) reported on the determination of zinc, cadmium, and mercury by fluorescence. Since then, good descriptions of the method have appeared in book articles by Vickers and Winefordner (100), Augusta Syty (101) (both these references contain long lists of detection limits), and Kirkbright and Sargent (102) and in numerous journal publications.

For the alkalies and the alkaline earths, whose resonance lines fall in the visible and infrared, flame emission is undoubtedly superior, but in the ultraviolet, where this method is weak or inapplicable, fluorescence is efficient. The two methods thus in a sense complement each other.

The question might be asked, Why, with all these attractions, hasn't atomic fluorescence for routine analysis been used at least to as great a degree as AAS? The answer, apparently, is the dearth of instrumentation. Very few absorption instruments in use today have the capability of conversion to the fluorescence mode, and furthermore, the building of filter photometers saves only the cost of the monochromator, a cost that appears to be too small to be attractive to manufacturers. However, at present a vigorous research program is in progress by workers interested in methods development. The routine worker, whose interest is simply to get the analytical job done, and who may not have the necessary skill in instrument design or even the needed shop facilities, must depend on commercial availability.

3.9   MULTIELEMENT ANALYSIS

The question of multielement spectroscopic analysis by atomic absorption, emission, and fluorescence has received a thorough review by Winefordner and co-workers (103). Atomizing devices discussed are flame, inductively coupled, and microwave plasmas, with particular applicability to multielement analysis. Detection devices (temporal, spacial, and multiplex) are considered.

The authors concluded that AAS is unsuited for this application, but that flame and fluorescence can be adapted, with certain changes in the monochromator, the atomizer, and the exciting beam. Here again the practical analyst must await action by instrument designers and researchers to apply these new ideas, but this is undoubtedly the trend for the future.

Lists of manufacturers and suppliers of atomic spectroscopic equipment have been published by Dean (104) and Veillon (40). However, these lists, although comprehensive, are now 8 or 9 years old, which in this rapidly changing field makes them unreliable. The best source of current equipment is still manufacturers' brochures and catalogs. A short list of the principal suppliers in their respective fields will be found in the Appendix.

TECHNIQUE

AAS is a relative method: Unknowns are compared to standards, in contradistinction to absolute methods of which the gravimetric is the commonest example. Results in AAS can therefore be no better than the standards. In addition, the sample, to be meaningful, must truly represent a much larger mass of material, and this requires care and understanding of the sampling technique. The actual measurement is so easy and routine that the principal labor in the procedure is in preparation of the sample. Thus, the main effort should be in designing the overall procedure to make it as simple as possible and still give answers commensurate with the problem at hand.

If the standards are not too different from the unknowns, both in concentration and in composition, then confidence in the results can be fairly assumed. However, certain types of samples are difficult to match with respect to composition because they are so variable—for example, samples from soil and mineral surveys, air samples in environmental control, smears for dusts, and some agricultural and biological samples. Industrial materials, such as metals and chemicals, are much more uniform in composition and are thus capable of giving much more precise results in analysis. Materials that are items of commerce require the most care in analysis (more precisely, assay) because it establishes their monetary value. This kind of analysis is usually called umpire or referee analysis.

4.1 PREPARATION OF THE SAMPLE

4.1.1 SAMPLING TECHNIQUES

Variable materials such as rocks, ores, soils, sediments, and mining tailings and concentrates present some difficulty in sampling. Geological materials may be highly segregated, and soils and sediments may classify according to gravity because of water action. Any material flowing in a stream—for example, on a conveyor belt or in a slurry—is sampled according to the general rule of removing a portion of the flow all of the time or a whole cross section for some of the time of flow. Samples that end up in a large bulk of 50 lb or more are cut down by a process of coning and quartering. The material is piled in a cone, divided into four equal quarters, and two opposite quarters retained while the other two are rejected. The process is repeated, with crushing to a finer size if that is indicated, until a weight of 2 to 3 lb is obtained. This is pulverized and riffled down to about 5 g, which is then further ground until it all passes a 200 mesh sieve.

The equipment needed for such an operation is a laboratory jaw crusher, a disc pulverizer, and either a manual or mechanical mortar and pestle. Another device for the final grinding is the impact mill or Wigglebug. Sieves of brass or stainless steel will inevitably contaminate the sample; nylon sieves should be used if metal contamination is undesirable.

More detailed information can be obtained in Prior (105) and in publications listed in a bibliography of sampling methods (106). The ASTM has published a recommended method of sampling coal (107) that can be applied to similar materials. For silicate materials, the textbook by Hillebrand and Lundell (108) is particularly good.

A sample is a small mass of material representative of (in the sense of "standing in the place of") a larger mass. This is a statistical concept, whose interrelated factors are size of the final sample, fineness of comminution, and degree of segregation in the original material. A fourth factor is the analytical precision required, for if it is low the labor in sampling should be restricted to match the precision. Wilson (109) has made a mathematical study of the effects of these factors on the randomness of the final sample.

Metals and alloys are generally sampled by drilling. Using a mill, shaper, or file for this purpose will tend to favor material from the surface and not from a depth. Wright (110) discusses this subject, and it is also included in the ASTM publication on spectrochemical analysis (111).

The sampling of ecological materials (vegetable matter, waters, and airborne dusts) was reviewed by Calder in 1964 (112), and Allen (113) published a more comprehensive text in 1974. Jacobs (114) discusses the problems encountered in collecting and sampling industrial inorganic poisons.

Fibrous materials, such as foods, animal feeds, and vegetable fibers, can be comminuted in a Wiley mill. Fresh water, seawater, and sewage, which may contain undissolved solids, should first be filtered. Petroleum lubricants, such as engine oils, may contain metals both in suspension and in true solution in the form of metalloorganic compounds, and should get appropriate treatment, depending on what the analysis is intended to show.

Archeological and art objects, where the quantity of material must be at a minimum, are often sampled with a dentist's burr; a low-power dissecting microscope is efficient for picking out individual grains from a heterogeneous mass.

Sampling problems are diverse; not only do errors arising here render the analysis useless, but the deceptive results can cause great harm. It is unsafe to assume that correct sampling technique is so obvious that it can be left to untrained persons. The chemist responsible for setting up an analytical procedure must also be responsible for sampling.

4.1.2 DISSOLUTION METHODS

Metals and alloys that are to be put in solution in acids must first be put in a suitable form, by drilling, by chipping in a shaper or miller, by turning in a lathe, or by filing, but contamination by the cutting tool is always a danger. The contaminating metals, in addition to iron, are the common ferrous alloying metals. Most metals yield to a hydrochloric acid treatment; more resistant metals must be dissolved in aqua regia. Some elementary precautions must be remembered. Nitric acid will precipitate tin, and sulfuric acid will do the same to lead, strontium, and barium. Spectrographic examination of the common acids indicates that they are generally prepared in stainless steel vessels, because traces of iron, manganese, nickel, and cobalt are found; if these metals are to be determined in the sample then blanks must be run. In fact, blanks should always be run, even when Utrix acids have been used.

Silicates are customarily broken up by a sodium carbonate or sodium peroxide fusion, the former in a platinum crucible and the latter in a nickel crucible. Platinum or nickel will then appear in the solution, and some easily reduced metals, such as lead and tin, may alloy with the metal of the crucible to a slight degree and contaminate succeeding samples. Some silicates may be broken up more efficiently by hydrofluoric and sulfuric acids, followed by fuming to SO_3 fumes, to get rid of HF and silica. The classic textbook by Hillebrand and Lundell (108) is an excellent guide to these treatments of silicates. A monograph on decomposition techniques of inorganic materials has been published by Dolezal et al. (115).

Bernas (116) describes a pressure vessel of Teflon, with the mineral acids as solvents, as a substitute for fusions with their additions of large excesses of salts to the sample. This Teflon autoclave is now available commercially. Recent accounts of its use appear in references 117, 118, and 119.

In samples containing much organic matter, the carbon must first be destroyed. This problem of removing organic matter is so pervasive in analytical chemistry—arising in agricultural, biological, ecological, forensic, and industrial analysis—that Gorsuch has written a monograph on the subject (120). The general method of getting rid of organic matter is by dry-ashing or wet-ashing. Neither method is entirely satisfactory, but there really is no better.

Dry-ashing is simple and direct. The material is heated in a platinum, silica, or porcelain vessel to at least 500 to 600°C until all black particles are gone. Mercury, cadmium, and zinc are also driven off and lost. Partially lost are nickel, vanadium, and lead. Moreover, this treatment often results in a residue that is difficult to dissolve in acid, so that it must be fused in sodium carbonate, followed by precipitation or solvent extraction to remove the large quantity of sodium salts in order to avoid plugging the burner slot. Impurities from the fusion and the acid invariably appear in the final solution (121). Certain samples are unsuitable for dry-ashing. Blood tends to foam on heating, some industrial samples spatter badly, and plastics melt to a viscous mass.

Wet-ashing has its own problems. The organic matter is destroyed by strong oxidizing reagents—aqua regia and sulfuric acid, with perchloric acid replacing the sulfuric if lead, barium, strontium, or much calcium is contained in the sample. The digestion is generally carried out in an Erlenmeyer flask covered by a short-stemmed funnel to act as a reflux condenser. The flask contents are gently boiled until a colorless or pale yellow solution results (122, 123). This is a slow process.

Wet-ashing causes no loss of inorganics, but contamination results from the large quantity of acid used and from leaching of the vessel walls. If perchloric acid is used, an explosion may result from the least carelessness. Samples containing glycerol, alcohol, or esters should not be used with perchloric acid, and in general nitric acid should be added to the flask just before the fuming stage, as a safety measure. Perchloric acid should be treated with the utmost respect.

Leaves, stems, bark, and other bulky samples are clumsy to handle by wet-ashing. If volatile elements are to be conserved, then a closed system should be used (124, 125).

Perhaps the best way to avoid the problem of loss when dealing with the volatile elements is to do away with preparation altogether. Typical of this course in recent years is the paper by Wright and Riner (126), who described the determination of cadmium in blood and urine by diluting the samples with water and placing a measured volume in a graphite furnace, where, after careful programming of the temperature, the cadmium was atomized. The furnace method can be applied to many other types of samples.

Oils and greases present a special problem (127–129). In oils the cations are frequently combined as metalloorganic compounds that are difficult to break up. Instead, they may be diluted in any suitable solvent and aspirated directly into the flame for total metal content. Suspended particles do not appear to interfere. See, for example, the paper by Miller et al. (130) for an account of their experiences. Free and combined metal may be distinguished by filtration and determined separately.

4.1.3   CHEMICAL PREPARATION

Not every sample type needs a preliminary chemical preparation, but for those that do, the purposes can be listed as follows:

1. To concentrate the element of interest to a level that is convenient for its determination.
2. To remove the element of interest from the bulk of the matrix so the atomizer can handle the solution.
3. To remove the element of interest from interfering materials.

Several techniques can be followed in preparing the sample. While it is true that they add to the labor and time of an analysis, the procedures are not complicated and they do greatly extend the versatility of AAS.

Organic Solvents and Extraction

The most generally useful chemical method of sample preparation for atomic absorption is to extract metal from an aqueous solution into an organic solvent. This can be done either by taking advantage of the greater solubility of certain inorganic salts in organic solvents or by chelating the element of interest and extracting the chelate into the organic solvent. One advantage of solvent extraction is that sensitivity is enhanced in the flame by a factor of 300 to 500% when an organic solvent replaces an aqueous solution.

Chelation and extraction methods were studied first by physical chemists, who worked out the underlying theory. Applications to problems in analytical chemistry have been published by Morrison and Freiser (131), Stary (132), and many others. Dean and Lady (133) were the first to use extraction methods in flame photometry. Allan (134) studied the technique in conjunction with atomic absorption and found that esters and ketones were the most satisfactory organic solvents. He also pointed out that the most generally useful chelating agent for atomic absorption should be unspecific and noncritical with respect to pH so that a simple chemical procedure can apply to a number of elements. The burden of specificity should fall entirely on the atomic absorption method, thus providing a total technique that is free of interferences and yet is quick and convenient.

The operation consists in adding the chelating agent to a measured volume of the aqueous sample, adding a measured volume of the solvent, separating the two phases in a separatory funnel, and then nebulizing the solvent directly. The solvent should not be soluble in water (to avoid troubles arising from emulsification and loss of analytical volume), and it should behave well in the burner: there should be no smoke, and it should be possible to adjust the gas flows to achieve a stoichiometric flame of the burning solvent. Chlorinated solvents give rise to toxic combustion products and should be used only with a good ventilating system. They also burn badly. Suitable solvents, as Allan reports, are esters and ketones; of the latter, MIBK (methyl isobutyl ketone) is widely used.

Of the chelating agents, APDC (ammonium pyrrolidine dithiocarbamate) is currently very popular because it is broadly applicable to the extraction of many of the heavy metals from acid solution. It is possible to perform the extraction, in most cases, at a pH near 2.5. A recently compiled list from many sources of chelating agents and solvents for specific metals can be found in Kirkbright and Sargent (135). Another recent publication on the subject is by Lemonds and McClellan (136).

APDC can be obtained commercially but can also be prepared in the laboratory according to a procedure described by Malissa and Schoeffmann (137). Many workers prefer to prepare fresh APDC solutions each day because of the instability of the solution; the need for this, however, has not been established.

Precipitation

Another powerful method of separation and concentration is by chemical precipitation, the classic technique of analytical chemists before the advent of instrumental analysis. By now a wealth of information, together with descriptions of thoroughly tested procedures, can be found in numerous textbooks on gravimetric and volumetric analysis. Some easily available books are listed in references 138 to 141.

The power of the precipitation process is evident when the analyte must be séparated from a solution containing a high concentration of salts, which might interfere chemically or clog the burner head. The common example is the acid solution of a sodium carbonate fusion. This type of separation can be combined with a concentration of the analyte into a much smaller volume.

Microgram quantities of analyte can be handled by the process of coprecipitation, in which a small amount of an agent that acts as a collector is added to the sample solution and the two are precipitated simultaneously. A common example of this technique is the precipitation by H_2S in acid solution of such metals as lead, cadmium, copper, silver, and bismuth, where one of the metals can act as collector for any of the others. For the transition metals, a common collector is aluminum with ammonia as the precipitant. Bulky, flocculent agents appear to be the best collectors.

Preparation of the sample for AAS is much easier than for gravimetric procedures. Cleanliness of the precipitate is of very minor, if any, importance. In AAS the monochromator distinguishes between the analyte and any impurity, and the detector-readout does the quantitative measurement. Thus, no thorough washing and no double precipitation are needed.

Removal of filter paper presents little difficulty. It can be destroyed by wet-ashing or by ignition, followed by solution in acid. Or, still easier, the precipitate can be washed from the paper while it is still in the funnel.

Certain organic compounds, familiar to all analytical chemists, have long been used as precipitants for metals. Examples are cupferron, 8-hydroxyquinoline, dimethylglyoxime and several others. These are more specific than either H_2S or ammonia. Precipitates of these compounds can be redissolved into an aqueous solution, or into an organic solvent and aspirated directly. The book by Lundell and Hoffman (142) is arranged in a particularly convenient manner to show solubilities in various menstruums. Also see, for example, the article by Mitchell (143) on precipitation concentration of elements in biological and plant material.

When samples cannot be measured directly but must be changed in concentration, in matrix, or for any other reason, the precipitation technique should have first consideration.

Ion Exchange

The term ion exchange refers to the exchange of ions (for our purposes, usually cations) between a solution and an insoluble solid. To be practical, the solid must be permeable so the solution can make good contact within the body of the solid and thus produce a rapid exchange. The solids for analytical use are synthetic organic resins. A familiar application of ion exchangers is in water softeners.

The technique in applying ion exchangers for AAS is simple and takes the form of either a column or batch operation. In the former, a buret (or commercial column) is packed with the appropriate resin and the sample solution is passed through it. The elements of interest are adsorbed onto the resin and then recovered by treatment with a strong acid. In a batch operation, the sample solution is simply mixed with the resin in a suitable container and then removed by filtration. Treatment of the resin with an acid then removes the wanted elements.

The exchange process is capable of some concentration if the final volume is less than the original, and of course it can be concentrated further. There is always danger of contamination. Batch operation is quicker and is used when a group of elements is to be separated from a matrix.

Literature on the ion-exchange process is extensive, first developed by workers in chromatography in the 1950s. Some textbooks treating the subject are listed in references 144 to 148, and some recent applications to problems in atomic absorption are in references 149 to 153.

It should be noted, however, that ion exchange is slow, and the passage of solution through the column must be watched so that none of the resin is allowed to become dry.

Other Concentration Methods

Simple evaporation to remove the bulk of solvent is a method older than chemistry itself. It has been used with natural waters for the concentration of dissolved solids. Even mercury can be evaporated (under a hood!) slowly to remove bulk and concentrate amalgamated metals. This property can be exploited when a mercury electrode has been used in an electrolytic separation of the wanted metals from a large volume of solution. Mercury can also be used as a direct solvent of certain metals, as with ores or minerals.

Volatilization can be applied to some elements. The removal of silicon in the form of volatile fluoride is a familiar example. Other elements for which volatilization is practicable are arsenic, selenium, germanium, tin, uranium, and nickel, after they have been converted to a volatile form. Minczewski (154) lists methods of separation by volatilization for many elements.

Fire assay is another method, old but effective, for concentrating gold, silver, tin, and the platinum metals from their ores and minerals. The process consists in fusing the sample with PbO (litharge) and a reducing agent such as flour, starch, or sugar, and then driving off the reduced lead button by cupellation, leaving a bead containing the alloyed metals. The metals can then be put in solution and aspirated into the flame. Fire assay requires considerable experience and is hardly justified for single element determination, but the method is capable of great concentration and is very effective for the collection of nine metals in one operation.

Another separation-concentration technique, although of limited use, is electrodeposition. It is useful only for concentrating the wanted metals from the matrix, not for removing the matrix from a trace of metal, because the metals in solution tend to adsorb on the plated-out metal. Rollin (155) and Rogers (156) discuss electrolytic separations, and Ashley (157) has published a review of the subject.

4.1.4   SPECIAL TREATMENTS

This section deals with methods in which special treatments are needed, because of either the smallness of the total sample or the need for maximum precision, as in the case of umpire analysis, or for other reasons.

Trace Analysis by the Delves Cup

The Delves cup technique, which should appropriately be entered here, has been discussed in Section 3.7 in connection with a description of its equipment. The method is suitable for the determination of the more volatile elements like lead, cadmium, thallium, and zinc; it requires only a small sample, about 100 μl or less; and it is more sensitive than conventional flame AAS. Hitherto it has been widely used for the determination of lead in blood, but for this problem it is now being superseded by the graphite furnace.

Atomization by the Electric Furnace

The idea of an electric resistance furnace for the atomization of samples originated with the Russian B. V. L'vov (158). The detection limits he obtained by this method of volatilization were so great an improvement over conventional flame technique that his report aroused immediate attention. His results for some common metals, based on the weight of element to cause 1% absorption, were:

Zn	8×10^{-13} g	Tl	4×10^{-11} g
Cd	2×10^{-12} g	Mg	5×10^{-13} g
Ag	8×10^{-13} g	Cu	1×10^{-11} g
Fe	2×10^{-11} g		

These detection limits are lower than flame results by two to three orders of magnitude. This is not unexpected, for furnace atomization uses a very small sample, 10 to 100 μl, and peak absorption is reached, and measured, in milliseconds, whereas for the flame the sample volume is much larger and readings are taken at steady state. Furnace detection limits of 10^{-11} to 10^{-13} g compete directly with limits reached by neutron activation and mass spectroscopy, which use equipment very much more complex and expensive.

Several years after L'vov's first paper, the inevitable improvements in furnace design appeared. Woodriff and colleagues (159–161) suggested a much longer cuvet. In the same year, Massmann (80) proposed a simpler version of the L'vov design. The latter two designs, however, produce about the same detection limits as the original. At present, much work is being done in the design and application of furnaces.

Furnace atomization should be looked on as a specialized technique where sample size is limited and the analyte is present in trace quantities. Where these conditions do not put a restraint on the procedure, flame atomization is preferable, because the routine is quicker and the precision better. Furthermore, the flame technique needs no such expensive accessories as deuterium background corrector, temperature programmer, and more elaborate readout system.

A determination by the furnace requires about 4 min for the full cycle, and reproducibility is only about 6% on average. One attempt to improve both these parameters is an automatic pipetter, described by Slavin (162), which is now available commercially. This device can transfer 10 or 20 μl of sample aliquots into a cuvet from each of 30 stations mounted on a circular table. The pipet is automatically rinsed between samples. Reproducibility is claimed to be about 0.4%, and precision with real samples about 4.0%. Figure 4.1 is a recorder trace of repetitive runs with the automatic pipetter.

L'vov, in his textbook on AAS, remarked that an improvement in precision and, more important, the elimination of most interferences could be achieved by integrating the furnace signal over time rather than using the peak height reading. This process requires a rapid-scan recorder or oscilloscope and some calculation, making it impractical, but at least one manufacturer has designed a readout for this type of integrated reading. It operates on the principle of summing up the signal at intervals of 0.2 sec.

Another essay in making furnace operation more convenient is that of Hwang et al. (81). During the temperature-programming cycle, the furnace is tied up for the drying and charring steps, which can more easily be done away from the furnace. Hwang et al. attacked the problem by placing the sample in a graphite boat, treating it in batches on a hotplate, and then inserting it into a special cuvet (Fig. 4.2).

Another idea of extending applications of the furnace is that of Langmyhr and co-workers (163–167), who investigated the possibility of

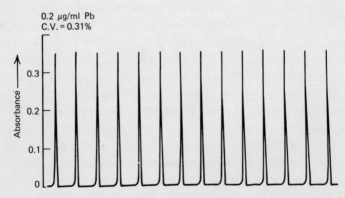

Fig. 4.1 Trace of 14 samples pipetted automatically into a furnace cuvet. Note the consistent results.

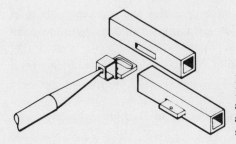

Fig. 4.2 Microboat antomizer system of Hwang et al. (81). The microboats are about $10\times8\times1$ mm in size and can hold about 80 μl of solution. Courtesy of Instrumentation Laboratory, Inc.

atomizing solid samples. Although the graphite boat would have been more convenient for their experiments, they used conventional cuvets, exposing one end and introducing the sample into the tube by means of a micro spoon. The samples were of such diverse materials as ferrosilicon, ferromanganese, paper and pulp, dental alloys, and various silicates. Elements determined were cadmium, lead, thallium, gallium, indium, silver, and zinc (note that these metals are all comparatively easily vaporized). Samples were weighed on a micro balance and were limited to 25 mg or less.

The precision obtained by the Langmyhr group varied greatly, which was to be expected because of the variation in degree of segregation in the small samples. Wilson (168) pointed out some years ago, in a mathematical study of segregational errors, that these errors are a function of the sample size, the degree of comminution, and the fineness of the different particles in the original material.

A late review of furnace construction and operation has been published by Woodriff (169). Some ingenious applications of the furnace are described in the papers summarized in Chapter 6.

In a study of the chemistry of atomization, Campbell and Ottaway (170) found that chlorides tend to vaporize, at least partly, in the molecular state—that is, with unbroken bonds—and therefore to that degree do not add to the absorption signal. These workers suggest that wherever possible the sample should be in the form of the sulfate or nitrate, which would produce more free atoms.

Another chemical effect noted by workers is caused by the form of the cuvet graphite. With ordinary graphite, the cuvet wall is permeable to metallic vapors, which are lost to absorption. Another fault is the tendency of certain elements to form very refractory carbides, which cannot be completely removed during the high-temperature cleaning stage. They then cause a loss for the first sample of a run and memory effects in succeeding samples. These troubles are largely cured by pyrolizing the inner graphite surface of the cuvet. Siemer, Woodriff, and Watne (171) describe a simple laboratory technique for coating graphite tubes with a pyrolitic layer.

Aspiration of Small Volumes of Solution

Sabastiani et al. (172) sought to determine how small a volume of sample solution in the conventional flame burner-nebulizer system would still produce the same sensitivity as the usual 2 ml sample volume. They found this minimum volume to be 100 μl. Their procedure was to suck up the measured globule from a plastic surface (which was not wetted) by means of a capillary tube feeding into the nebulizer. In the course of their experiments they noted that solutions of high salt content did not clog the burner head because of the small volume used.

Manning (173) took up the idea but changed the method of injection slightly. He used a small conical cup of Teflon (Fig. 4.3) attached rigidly to the nebulizer intake, a short-period recorder, and solution volumes of 100 μl. He found that precision was not quite as good as that obtained with the

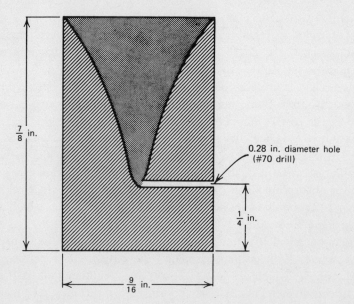

Fig. 4.3 Conical cup for injecting a small sample into a conventional nebulizer-burner.

usual large volume, although sensitivity on an absolute basis was improved by a factor of 20. Compared to furnace atomization, sensitivity was inferior but precision was better. For these two criteria, therefore, the small-volume technique was half-way between the two standard techniques. However, it has advantages, which may be listed as follows:

1. The standard flame and burner setup may be used.
2. Solutions with high salt content may be treated without fear of clogging the burner head.
3. Extremely rapid throughput can be achieved.
4. Preparation of the working solution can be easier.
5. Cost of the conical cup is trifling.
6. Viscuous or difficult samples, such as serums, can be run.

Manning demonstrated the first of these advantages by analyzing plant tissue. The sample was only 0.3 g of NBS SRM 1571 (orchard leaves), from which 6 ml of solution was prepared. This served to determine 11 trace elements in duplicate, with generally good agreement with the certificate values.

As another example, a sample of blood was analyzed for copper and zinc, with relative standard deviations of about 3%. Each element was determined 16 times, and the replicates required only about 2 min to complete. Contents of copper and zinc were unknown, but a comparison run of an aqueous solution containing 1 μg/ml of the two metals indicated that much lower concentrations could be determined easily.

Jackwerth and co-workers (174–176) used the small-volume technique, which they called the "injection method," to determine trace impurities in real samples. Their procedure was to start with a large sample and concentrate the impurities in a small volume by precipitation and resolution. Their samples were alkali and alkaline earths, high-purity gallium, and manganese compounds. Limits of detection were found to be between 0.06 and 0.003 ppm, and relative standard deviation was generally about 4%.

Experience with the small-volume technique is so recent and so sparse that its value cannot yet be assessed. Manning modestly calls the technique a "complement to, rather than a replacement for, any present sampling method." Already an improvement has been suggested by Goulden (177). He extended the time of sample flow by controlled dilution with air, thereby doing away with the need for a rapid-response readout and reducing the minimum volume to 25 μl, and still obtained precision comparable to macrovolume techniques. With further experience, the small-volume technique may become routine, largely replacing the elaborate and expensive furnace atomizer, particularly where ultimate sensitivity is not needed.

The conical injector vessel is easily made where machine shop facilities are available, or it can be bought from Perkin-Elmer.

High Precision Analysis

When considering the question of high precision analysis, we may simplify the problem somewhat by ignoring the factors that don't apply. We may assume that high precision methods apply only to major constituent analysis, for at trace concentrations precision falls off badly, and generally it is not of interest here. In addition, we assume that plenty of sample is available and therefore the flame is the choice over the furnace.

With ordinary care and basic instrumentation, routine procedures will produce a precision of about 1 to 2%, as measured by the relative standard deviation (RSD), when working with concentrations that are low but much above the detection limit. But with high-concentration samples, various expedients can be tried to increase this precision by a factor of three to five. For matching compositions of standards and unknowns, the method of standard additions can be used. The unknown should be bracketed closely between standards. Scale expansion and a digital readout will reduce scale reading errors. Readings can be made to fall on the most favorable portion of the absorbance range, and the equipment should be capable of linearizing the working curve and also of integrating the signal over time. A double-beam system will guard against drift of lamp output.

The good precision possible with the AAS method was recognized early in its development. In 1965, Meddings and Kaiser (178) found that AAS procedures at 1% RSD were quite good enough for routine plant control analyses but not quite the equal of gravimetric procedures for assays involved in the purchase of ores for a custom smelter. For these, an RSD of about 0.1% is required. However, they failed to mention that this precision level is exceptional; many gravimetric analyses cannot be done to this precision.

Weir and Kofluk (179), working in the same laboratory and at the same time, studied sources of indeterminate error in the AAS procedure. Examining volumetric apparatus, they found that precision in containment in, and delivery from, flasks and pipets ranged from about 0.04 to 0.16% RSD. Weighing operations at the 1 g level produced an error of about 0.01% RSD. Dilution operations produced an error of about 0.1% RSD. Errors in reading burets depended strongly on the experience of the operator.

Two examples of high precision analysis for major constituents are given in more recent publications. Thomarson and Price (180), working with ferrous alloys of the official British Standards series, compared their AAS results with certificate values. Their results fell within or very close to the certificate range from which the certified value was derived.

Price (181) published similar data, but on major constituents of various nonferrous alloys, comparing his AAS results with referee-grade analyses. He reported very good agreement with the gravimetric results and in his precision tests obtained an RSD of about 0.35%.

Working with the latest sophisticated apparatus in the applications laboratory of an instrument manufacturer, Fernandez and Kerber (182) sought to determine the limit of precision obtainable. They got somewhat better precision than the English workers on their test samples. These were several NBS Standard Reference Materials. Their agreement with certificate values approached in quality the most careful gravimetric work.

One of their test samples was NBS #1013, cement, with a certified CaO content of 64.26%. They obtained an average of 64.30% CaO on 30 repeats over 5 days, with a spread of 0.17 to 0.29 for the daily RSD.

One of the new accessories in this work was a microprocessor, which helped with respect to speed, convenience, and overall performance, especially the ability to calibrate automatically over a very wide absorbance range without reducing precision.

Ingle (183) took a more theoretical approach to the question of precision. He developed equations to indicate dependence of RSD in absorbance on various instrumental and chemical variables, and showed how they could optimize precision for a given problem. In a later paper (184), the theoretical predictions were checked by experiment—repetitive flame measurements on copper solutions. The study showed that under typical conditions at low concentrations, noise due to fluctuations in flame transmission was the limiting factor; at high concentrations, noise due to absorption properties of the analyte was the limiting factor.

From these examples we may conclude that with correct and careful

chemical preparation, well-designed equipment, and a thorough knowledge of its use, a precision of 0.2 to 0.3% RSD is attainable. This precision
level can make AAS competitive with high quality gravimetric analysis—
indeed, with the state of the art.

Indirect Analysis

Several of the nonmetals, whose ground state lines fall in the far-
ultraviolet, out of reach of the standard AAS instruments, can nevertheless be determined by indirect methods. One that readily comes to mind is
the precipitation of the SO_4 ion by an excess of barium chloride, followed by measurement of the residual barium. The slight solubility of
$BaSO_4$ can be corrected for by using standards through the same procedure as for the unknowns. Varley and Chin (185) report on this method for
sulfur determination.

Another obvious application of the indirect method is the measurement
of chlorine, bromine, and iodine by precipitation with silver nitrate and
measurement of the excess silver by conventional AAS, as described in a
paper by Truscott (186). In fact, this idea can be applied to any element
that can be separated, by any convenient means, from an agent with
which it is tied stoichiometrically with a determinable element. Methods
have been reported for the halides and sulfur, phosphorus, carbon, and
nitrogen.

Another group of elements, which are difficult to determine conventionally because their sensitive lines are in an unfavorable region, or
because they present difficulties in atomization, can be treated by complex reactions with phosphomolybdic acid, followed by determination of
molybdenum. Danchik and Boltz report on such a procedure for arsenic
(187), and Jacubiec and Boltz (188) determine germanium as the molybdate. See also pages 161–162.

Isotope Analysis by AAS

Determination of isotopes is traditionally done by mass spectroscopy.
The cheapness, simplicity, and rapidity of AAS and its equipment have
led workers to attempt this analysis by atomic absorption. The principal
problem in isotope analysis is the high spectroscopic resolution needed to
separate the extremely close lines of most isotopic mixtures. High resolution in itself is not too great an obstacle, as we have large echelle spectro-

graphs capable of the resolution needed, but the radiation energy transmitted by these instruments is small, and besides the equipment and procedure are no longer simple, cheap, or rapid. For most elements, mass spectroscopy must still be the preferred method.

Of the elements that may be practicable, only lithium and uranium appear promising because their isotopes are unusually well separated. But they still require specially modified equipment, such as watercooling of the exciting lamp to produce maximum sharpness of lines. Goleb (189) has worked with uranium, and more recently Rossi (190) investigated the isotopic shift of several lines in the uranium spectrum. Lithium isotope analysis has been investigated by Wheat (191).

One hope for the solution of the isotope problem lies in the development of tunable lasers. Research in this field is proceeding at a great rate. Spectral widths of laser radiation are inherently so narrow that they should provide the needed resolution. The function of the conventional monochromator will then be reduced to excluding interfering lines.

4.2 INTERFERENCES

The term "interference" may be defined as that special effect which causes a determination to be in error. Factors affecting the precision or sensitivity of the determination are not considered interferences. The great attraction of atomic absorption over other spectroscopic methods is its relative freedom from interferences.

The subject of interference is still in a confused state. The literature is replete with papers on the theme—many of them contradictory—that endeavor to explain the causes and cure of errors. For this reason, the standard procedure of consulting the literature when having trouble may not be very effective. Errors due to interferences as defined here can be of three sorts: chemical, physical, and spectroscopic. In the first two categories in particular, many of the errors can be ascribed to a poor match in gross composition between standards and unknowns, so first consideration should be given to the composition problem. This may not always be under the control of the operator because of such factors as limited sample quantity or great diversity of sample composition. Diversity, however, can be overcome by a chemical separation, such as precipitation with a collector, at the cost of some time and labor.

4.2.1   CHEMICAL INTERFERENCES

In considering how chemical interferences play a role in flame spectroscopy, it is useful to recall the steps required to convert the sample solution
into an atomic vapor. When the premix burner is used for this purpose,
the nebulizer first converts the solution into an aerosol, a fog. This aerosol
is swept into the burner and thence into the flame. In the flame, the
droplets must first be dried, the residue melted and vaporized, and any
compounds dissociated to free atoms for absorption to be observed. If the
dried salt happens to be a compound that does not readily dissociate at the
temperature of the flame, the amount of metal observed by absorption will
be less than if its salt were easily dissociated. Thus, the presence of
phosphate in a solution of strontium will result in the formation of a
refractory strontium salt. It is known that the addition of certain cations to
a solution that contains an interfering anion will remove the interference.

A very clear explanation of chemical interferences and their control by
the addition of particular cations is given by Yofe, Avni, and Stiller (192).
If lanthanum is added to the solution containing strontium and phosphate,
the interference disappears. These authors found that lanthanum phosphate precipitates before strontium phosphate as water is evaporated
from a solution containing both. Thus, if sufficient lanthanum is added to a
solution containing strontium and phosphate, it will precipitate the phosphate as the aerosol is heated in the flame and permit the strontium to
precipitate as a more easily dissociated compound, say the chloride.
Other workers have corroborated this explanation by observing that there
is stoichiometric equivalence between an added suppressing agent and an
offending anionic complex.

The alkaline earths are particularly prone to chemical interferences in
the air–acetylene flame. In general, the interferences are totally controlled
by making up samples and standards in solutions of strontium (for calcium
and magnesium) or lanthanum (for calcium, magnesium, strontium, and
barium). Since lanthanum is rarely an analyte, its use is favored as a
protecting agent for the alkaline earths. Lanthanum oxide is available, with
only a small blank for the alkaline earths, from several suppliers.

Chemical interferences are a potential problem for most of the metals
that require a fuel-rich air–acetylene flame. Problems have been reported
for molybdenum, tin, and chromium. Workers have reported chemical
interferences on the determination of manganese and iron when low-
temperature flames are used, but only a silicate interference is observed
when stoichiometric air–acetylene flames are used for these two metals.

4.2.2 PHYSICAL INTERFERENCES

Most physical interferences occur in the nebulization process and have to do with such factors as viscosity of the solution and drop size. Viscosity directly affects the rate at which sample is being taken up and thus the concentration of analyte atoms in the flame. Drop size affects the rate of drying and of volatilization. Temperature of the carrying gas also has some effect on rate of takeup. The remedy for all these effects is to match the gross composition of standards and samples.

Most workers who have reported on the practical side of analysis have found it necessary to make this matrix match to about 1%. While the match is usually made by equalizing the concentration of major constituents, it is clear that this is a bulk or nonspecific interference resulting from physical changes of the solution. Another way of solving the problem of sample physical difference is to remove the analyte entirely (as by solvent extraction, ion exchange, precipitation, etc.), which adds to the labor of a determination but eliminates errors due to physical interference.

Light-scattering and flame absorption can also be classed as physical effects. Willis (193), in his early work, found that when urine aspirated directly into the flame contains none of the metal being sought, an apparent absorption occurs. He ascribed this to the scatter of light from the light beam by the small particles of salts that still survived as solid or liquid particles. The effect is particularly severe at the shorter wave lengths. A correction can be made by measuring a line of the sample close to the unknown line, but one not subject to metal absorption. This absorption is then subtracted from the unknown's and the correction used in plotting the working curve. Billings (194) has discussed this effect in some detail, and Barras and Helwig (195) have studied the magnitude of light scatter.

Scatter by the flame itself is not, strictly speaking, an interference, but may be mistaken for one. A fuel-rich flame scatters some radiation, even at the visible wavelengths. Since this takes place equally for sample and standards, it is canceled out in the process of setting the zero absorption control. All flames absorb at least a measurable amount of radiation at short wavelengths; this again is canceled out in setting the zero control. Absorption in addition to this is caused by organic solvents aspirated into the flame, especially below 250 nm. Again, this must be corrected for by subtracting the signal for solvent alone or by setting the zero control while aspirating pure solvent.

Another cause of interference is molecular absorption, owing to the formation of molecular species in the flame that are not dissociated. A common example is the presence of a high concentration of calcium in the sample. The temperature of an air–acetylene flame is too low to dissociate any calcium oxide formed in burning such a sample, and the oxide will absorb strongly at the resonance line of barium, as shown by Koirtyohann and Pickett (196). If barium is to be determined in a calcium matrix, variations in the calcium content will alter the apparent barium content. This effect disappears when a nitrous oxide–acetylene flame is used in place of air–acetylene, owing to the higher temperature of the former. Other similar effects, particularly in flames of lower temperature, are discussed by Elwell and Gidley (197).

4.2.3 SPECTROSCOPIC INTERFERENCES

Ionization, the removal of the valence electron from the influence of the nucleus, is the principal spectroscopic interference. Obviously, if the electron is not in the ground state when the scanning beam encounters it, absorption cannot occur. The fraction of the total atoms in the ionized state, therefore, will not contribute to the absorbance signal. The electron, or another, will eventually be recaptured by the nucleus, but by that time it may not be in the scanning beam's path.

The readiness of the atom to ionize depends on the temperature and on the atom's ionization potential, which, for those elements applicable to determination by AAS, ranges from 3.9 eV for cesium, the most easily ionized, to about 10.5 eV for the most difficultly ionized. This energy is supplied by the flame gases, and the relationship between ionization potential and temperature is given by Saha's law, which can be used to calculate the proportion of atoms in the ionized state at equilibrium for a specific element. Flame temperatures, although comparatively low, are nevertheless sufficient to cause appreciable ionization for many of the elements. Manning and Capacho-Delgado (198) tested Saha's law against experiment and found good agreement.

As ionization adds a factor of uncertainty to the measurement of absorbance, it can be minimized by matching sample composition or, more conveniently, by adding an alkali salt to all solutions, in an amount to bring absorbance to a "plateau" as shown in Fig. 4.4 for three of the alkaline earth metals.

Another, less common, form of spectroscopic interference occurs when

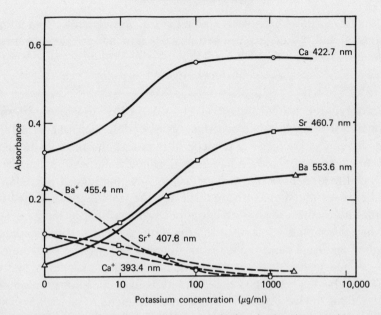

Fig. 4.4 Absorbance of calcium, strontium, and barium in the nitrous oxide–acetylene flame. Note the changes in absorbance with increasing concentration of added potassium salt for both the neutral and ion resonance lines of the three elements.

either a nonabsorbing line of the fill gas or a high-level line of the analyte element is unresolved by the monochromator at the setting of the analytical line. This resolution is, of course, affected by the slit width, so this is an additional variable. The net effect of this type of interference is a bent calibration curve and lower sensitivity.

If the interfering line originates in the fill gas, it can be eliminated by judicious choice of fill gas by the lamp manufacturer. If the interfering line is from the analyte element, another analytical line can be chosen, or both types of interference may be prevented from reaching the detector by narrowing the slit (reducing the bandpass). This, however, increases the gain required and thus the noise in the system.

The problem is augmented when using a multielement lamp, particularly if the lamp metals have complex spectra. A check may be made for the possible presence of interfering lines; this is done by scanning the region on either side of the analytical line wavelength. Additionally, a comprehensive atlas, like the MIT tables, should be consulted.

On the general subject of interferences, the literature is voluminous.

Some of it is contradictory, as stated before; consequently, testing must be careful and thorough when setting up a new analytical procedure.

4.3 ORGANIZATIONAL DETAILS

For a relative analytical method like AAS, where unknowns are to be compared to standards, the overriding consideration must be to maintain reproducibility. The general procedural plan and choice of instrumentation are imposed on certain nondiscretionary conditions like the size of the work load, sensitivity, and precision. A well-designed procedure is one that uses the simplest equipment and the least labor to do the job. Within these constraints, certain options are available to the analyst.

When the sample is plentiful and the required sensitivity is moderate, which is the usual case, the flame as atomizer is best. The measurement process and changing from one sample to another can be carried out rapidly. If the load is heavy, time can be saved by an automatic sampler. The readout device can be a simple microammeter. For high-precision analysis, the flame is better than flameless atomization.

The furnace atomizer provides certain advantages aside from its much greater sensitivity. A sample that cannot be aspirated smoothly because of its high salt content can often be diluted and treated in the furnace, thus avoiding a separation by ion exchange or organic extraction, and then followed by the flame technique. For samples of limited size or for solid samples, the furnace is a necessity.

However, there are certain disadvantages to the furnace atomizer. Equipment and the needed accessories are expensive. Precision in routine work is poorer, and operation slower, than with the flame. Because of the rapid evolution of the analyte, a simple meter must be replaced by a short-period recorder or a digital readout, although those meters with built-in peak-holding capability will also serve.

The method for aspirating small sample volumes, the injection method described previously, should also be considered. Although new, the technique is bound to be improved and may displace the conventional flame procedure.

4.3.1 ADJUSTMENT OF ANALYTICAL PARAMETERS

Measurements of absorbance should fall, whenever possible, within the optimum range of 0.2 to 0.7 absorbance unit. Degree of dilution of the sample can be determined approximately by consulting the data of Table 5.1 on page 112, which lists the analyte concentration needed to result in an absorbance of about 0.2 unit.

The wavelength position of the monochromator is set by peaking on the unobstructed analyte line emitted by the hollow-cathode lamp. The wavelength scale is used only as an approximate locator. There is always a danger that the wrong line has been chosen. Several checks against this error are available. The region on either side of the setting may be scanned; the lines emitted by the fill gas of the lamp should be well known; a standard of the analyte should be nebulized, and its absorbance should agree with the data of Table 5.1.

Adjustments of the nebulizer-burner assembly—gas mixture and pressure, height of flame with respect to optical axis, and aspiration rate—should all be such as to result in maximum signal. The gas mixture, whether oxidizing, reducing, or stoichiometric, can follow the recommendations in Table 5.1 or be determined by test. Burner adjustments are all explained in manufacturers' literature, which should be followed, at least at the beginning.

Composition of standards should be matched with unknowns. Failure to do this is a cause of errors, but matching is not always easy. However, several expedients are available. One is the method of standard additions, in which the unknown solution itself acts as a base for the series of graded standards. But this does not always work if unknowns are heterogeneous in composition. For this case, all may be brought to the same composition by applying one of the separation-concentration procedures outlined in Section 4.1.3.

On a negative note, the literature contains suggestions that an internal standard technique (199, 200), similar to the one commonly used in emission spectroscopy, could correct for compositional mismatch. But this is poor logic, because all sorts of unsuspected changes, both in the flame and in the wavelength setting, may then occur between measurements.

The range of standards should bracket the expected concentrations of the unknowns (no extrapolation), and where at all possible only a linear calibration curve should be used, although curve rectification in the readout can extend the calibration safely.

It must be remembered that metals in very dilute solution have a tendency to adsorb on the walls of the containing vessel and thus weaken the concentration with time. For long-time storage, concentrations of standards should be at least about 500 μg/ml and diluted only a day or two before use.

Normal procedure in AAS operation is to run standards before and after the series of unknowns, to verify that no shift has taken place. If the series of unknowns is long, an occasional standard should be run between unknowns. Blanks should be run also between samples, to check the stability of the zero setting.

4.3.2   SELECTION OF WAVELENGTH

The theoretical guide to choice of wavelength is the oscillator strength of the transition, usually expressed as the gf value. Practically, however, this guide is imperfect, for the gf values are known only approximately; the list is incomplete, and for some transitions a better line terminating somewhat above the ground state, depending on the flame temperature, may be more populated and therefore more absorbing. Also, interferences from close, nonabsorbing lines in either the analyte or the light beam may preclude the choice of the "best" line.

A recent article by Parsons, Smith, and McElfresh (201) thoroughly discusses this subject of line choice and should be consulted. An indispensable reference volume is the MIT table of wavelengths by Harrison (32). This is the most extensive compilation available and is basic for the location of possible interfering lines. Another important listing is the two-part NBS publication by Meggers et al. (202) which provides two listings of spectral line intensities. Meggers et al. also give the ionization potentials of the elements and the energy levels between which the transition for each wavelength occurs.

For certain elements the choice is limited to one or two lines falling in the working range of the usual monochromator, 200 to 800 nm. For elements with simple spectra the choice is the strongest line as listed in Table 5.1 of detection limits. These are the lines emission spectroscopists call ultimate or principal lines; a list of them can be found in the MIT tables. For elements with complex spectra, where the choice is not at all obvious, one can follow the experience shown in previous work or determine the suitable line by careful experiment.

4.3.3   OPTICAL ARRANGEMENTS

Whether the atomizer is a flame or a furnace, the hollow-cathode source is either imaged within the flame or cuvet or made parallel; the beam is then picked up by a second imaging unit and projected onto the slit. The projection system can be designed to use either curved mirrors or lenses. Mirrors have the advantage of being achromatic (all wavelengths fall on the same point), but their systems are more complicated. Lens systems are simpler, but for ultraviolet transmission lenses must be made of fused silica and, being chromatic, must be adjusted after more than small changes have been made in wavelength setting. One manufacturer (Instrumentation Laboratory) gets around this problem by using one or more zoom lenses in the projection system. With flames, a three-slot burner to make a wider flame may be used to ensure that the scanning beam passes wholly within the atomic vapor.

The function of the monochromator is to isolate the wanted radiation from extraneous light. A chopped beam does this too, but imperfectly, hence the need for a background corrector. This interfering radiation arises either in the atomic vapor or in the hollow-cathode lamp. In the former, the radiation consists mainly of molecular bands and some continuum. The possible presence of bands in the working wavelength can be checked by consulting the atlas of such bands by Pearse and Gaydon (203).

Interfering radiation from the lamp (atomic lines from the fill gas and the cathode metal) can be eliminated only by constricting the bandpass of the slit. If the identity of the interfering line is known, and the theoretical resolution of the monochromator is also known, then the possibility of separation by slit constriction becomes obvious. A well-designed monochromator should be capable of producing a working resolution no more than twice the theoretical. For maximum resolution, all grooves of the grating must be illuminated by the hollow-cathode beam. This can only be done by illuminating the slit with a converging beam at an angle equal to or greater than the angle subtended by the collimator at the slit.

4.3.4 SENSITIVITY AND DETECTION LIMITS

The terms sensitivity and detection limits occur frequently in the literature and are expressions used to denote the merits of an analytical procedure with respect to a particular element. The sensitivity is defined as the analyte concentration, in micrograms per milliliter, at which the absorption is 1% or 0.0044 absorbance units. It is usually determined by running a solution whose concentration is slightly above the expected sensitivity and then extrapolating down to the 1% level.

The defect in this figure of merit is that it takes no account of the noise level of the blank. This is illustrated in Fig. 4.5, which shows recorder traces near 1% absorption of both a noisy signal and a quiet one. Obviously, sensitivity for the latter is at a lower concentration than for the former, although both procedures by this criterion are equal. Although the test is to be run at optimum conditions for both, no account is taken of the sample composition or of the spectral region, each of which has a significant influence on background noise.

The detection limit, or limit of detection, on the other hand, is based on a statistical factor that does take account of noise at this low level of operation. Detection limit is defined as the concentration that can be "accepted with confidence as genuine and is not suspected to be only an accidentally high value of the blank measure." Quoted from IUPAC (204), this is as good a definition as any.

In practice this means that the concentration at the detection limit is to be stated within a specified number of standard deviations or sigmas. IUPAC recommends that this be 3σ; the significance of 3σ is that concentration at the detection limit would appear as a positive signal at a confidence level of 99.5%. Considering the difficulties involved in making this determination, agreement among workers (205–208) is surprisingly good.

Sabina Slavin and co-workers (209) have endeavored to reduce the uncertainty in determination of limits. They invoked modern electronics to integrate their readings over a 10 sec interval. This smoothed out a great deal of the background noise and enabled them to compare low-level operations of single- and double-beam instruments, in addition to constructing a new and more easily reproducible list of limits. This list has been included in the 1973 edition of the Perkin-Elmer "Cookbook" (210).

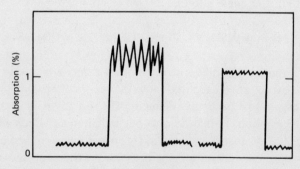

Fig. 4.5 Illustrating the difficulty of determining the sensitivity when the signal is noisy.

Also in the Cookbook is a list that takes an altogether different approach to the problem of presenting sensitivity data. This list tabulates the concentration of the elements that will produce an absorbance of 0.2 under standard conditions. To this writer this approach is more realistic and more practical; it has been reproduced in the present book as Table 5.1 (p. 112). At 0.2 absorbance unit, readings will usually still fall on the linear part of the working curve, and are still in the region where most analyses are made. The concentration datum can be established relatively precisely and so the figure of merit of some new experimental procedure can be conveniently compared to a standard, a valuable consideration. Too, from an absorbance of 0.2 to any other point is just as easy to calculate as from 0.0044 absorbance.

Furthermore, the old concept of detection limit has little practical utility. If such low levels are a concern in a given analytical problem, then it is better to make a preliminary concentration or to convert to the furnace technique.

4.4   THE CALIBRATION OR WORKING CURVE

4.4.1   THE CALCULATION OF ABSORBANCE

All light-absorbing methods are based on the Beer-Lambert law, which states, in AAS terms, that a light beam passing through a mass of absorbing gas decreases exponentially in intensity as the number of reacting atoms is summed up arithmetically. The mathematical expression of this law is usually stated as follows:

$$I_x = I_0 \times 10^{-KC}$$

where I_0 is the intensity of the entering beam, I_x the intensity of the emerging beam, C the concentration of atoms that the beam encounters, and K a constant encompassing such variables as path length and temperature. This equation may also be written in the form

$$\log \frac{I_0}{I_x} = KC$$

In this form the relation is a linear equation, making plotting and interpolation easy. With the older AAS readouts the practice has been to set the entering beam at 100 on the scale and at zero for the blank reading. I_x then measures the transmission through the flame, which is an inverse function of the quantity C we wish to measure. We obtain the direct function, the absorption, by substituting the quantity $100 - I_x$ for I_x. The left-hand side of the equation then becomes

$$\log \frac{I_0}{100 - I_x} = \log 100 - \log (100 - I_x)$$

$$= 2 - \log (\text{percent absorption})$$

This quantity $2 - \log$ (percent absorption) is called the absorbance, with A as the symbol.

The great advantage of plotting the absorbance against concentration is that it results in a straight-line calibration curve. This, however, holds for only a comparatively short range of concentration, about a decade or less for most elements. With increasing concentration, the calibration curve falls off from linearity, tending to curve downward toward the concentration axis. The effect is usually referred to as the failure of Beer's law; it represents a loss of efficiency of the absorption process and can result in increasing errors because of the reduced slope. To counter this, one must either dilute the solution or space the standards more closely. The best range of measurement, because of certain errors due to scale reading, which will be discussed below, is the range 0.2 to 0.7 absorbance unit.

However, it is not necessary to work in a narrow range and with dilute solutions to stay on the linear portion. Using a digital readout, electronic scale rectification, and especially time integration of the signal, it is quite practical to work on the curved portion and obtain sufficiently precise results, as demonstrated in reference 182.

4.4.2   THE CALIBRATION PLOT

Common practice is to plot absorbance on the y axis against concentration, in whatever units are chosen, on the x axis. Ordinarily, this results in a straight line passing through the origin of coordinates, provided the range is short. Extrapolation is unsafe as there is no telling where the curve will fall off from linearity. Rather, more standards should be run. The range can be extended considerably by the electronic curve-rectifying devices with which some AAS instruments are fitted. It is always good practice to repeat one or two standards at the high end of a series, as an insurance against curve shift.

Determination of Residual Impurities

The linearity of the curve and the arbitrary zero setting provide an important advantage in the preparation of certain standards. These are standards that, for the purpose of matching composition to unknowns, make use of reagent-grade chemicals with residual impurities. These impurities are often just the elements of analytical interest. Common examples are lithium associated with sodium salts, rubidium and cesium in potassium salts, calcium and magnesium in each other's compounds, hafnium in zirconium, titanium and niobium in tantalum compounds and metal, and copper and silver in supposedly pure mercury. In biological samples a similar problem of trace residuals occurs, with few reliable standards available.

The residuals can be evaluated by the method of standard additions. The compound can be spiked with a small amount of the analyte, and it and the unspiked sample run. In the plot of the result, the line passing through the two points cuts the y axis at some point above zero, and when extended to the x axis this intercept indicates the residual value sought.

A more accurate procedure is to make up the series of doped standards in the usual way and plot the results, as shown in Fig. 4.6. The curve projected into the minus x region intersects the concentration axis at minus 1.5 μg/ml; this is the concentration of the impurity sought. Or the

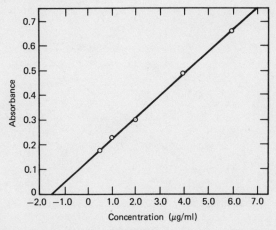

Fig. 4.6 Plot for the method of standard additions, shown here, indicates a residual analyte concentration of 1.5 μg/ml.

curve can be corrected by simply shifting to the right by 1.5 units or until the line goes through the origin.

The method of standard additions is a powerful procedure in the hands of the analyst. Consider this hypothetical case. Iron is to be determined in a complex industrial solution that contains three or four unknown major constituents and several minor ones. In the conventional way, a preliminary analysis would have to be made to identify and evaluate these constituents. A synthetic solution would then be prepared and doped with the iron standards; an iron blank on the reagents would still have to be made.

By the standard additions method, only the addition of a series of graded iron supplements is needed. This would have no effect on general composition, and no knowledge of this composition would be necessary; automatic matching would be achieved. Standard additions is a good general procedure to follow, provided enough sample solution is available to make up the standards and the concentration of the analyte and additions is such that the absorbance values fall on the linear portion of the calibration curve.

The problem illustrated in Fig. 4.6 can also be done algebraically. In Fig. 4.7 the point A represents a spiked standard and point B the unknown blank. We wish to determine the scale distance CO, the impurity. Com-

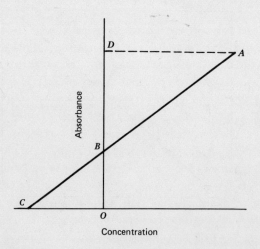

Fig. 4.7 Geometrical method of calculating the residual for standard additions.

plete the triangle *ABD* by drawing *DA* normal to *OB*. Then the triangles *ABD* and *CBO* are similar, so that

$$\frac{CO}{BO} = \frac{DA}{DB}$$

$$CO = \frac{DA \times BO}{DB} = \frac{DA \times BO}{DO - BO}$$

The conversion of these symbols into our practical quantities is obvious from an inspection of Fig. 4.6; that is, concentration of the unknown equals the spike concentration multiplied by absorbance of the unknown, divided by the difference in absorbances of the spiked and unknown samples.

4.5   STATISTICAL OPERATIONS

4.5.1   THE LINEAR CALIBRATION CURVE

The equation of a linear calibration plot is

$$y = mx + b \qquad \text{or} \qquad m = \frac{y - b}{x}$$

where y is the absorbance, x is the concentration, m is the slope, and b is the intercept on the absorbance axis. As the blank reading in AAS is adjusted to pass through the origin, b is zero. The slope, for all readings of the standards used to establish the calibration curve, is therefore simply $m = y/x$, or the absorbance divided by the concentration for each standard.

Thus, to establish from the standards data the best fit for the curve, it is only necessary to average the slopes. The curve can be drawn through the origin at this average slope. If it is felt that slopes determined at the lower concentrations are not as reliable as those for higher concentrations, then these latter can be given increased weight by, for example, doubling the middle concentration readings and quadrupling the readings at high concentrations in calculating the average.

Slope data are also used to determine the reliability of an analytical procedure, as explained in Section 4.5.2.

4.5.2　TESTS OF RELIABILITY

An analytical procedure worthy of the adjective "quantitative" should be such that the extent of errors to be expected is known. Errors fall into two categories—systematic and random. Systematic errors, by far the more dangerous, are the kind caused by some unknown or unsuspected factor. Common examples are impurities in reagents, partial solubility of precipitates, segregation, and mismatch of standards with unknowns.

Some of the techniques commonly employed to guard against such errors are the running of blanks, the check against calibration curve shifts, the use of doped samples (method of standard additions), the testing of separation techniques with known additions, and the run through of the entire analytical procedure with like samples whose composition is worthy of trust because of the great care with which it had been determined. These last are Certified Standard materials as issued by the National Bureau of Standards and similar agencies of Great Britain and Germany, by the U.S. Geological Survey (rock samples), and by various metallurgical firms (type alloys).

Random errors, on the other hand, are those that can go in either direction, either high or low, and, being random, tend to balance out. Examples of this category are the noise caused by instability in flame and electronic components, and small divergences in weighing and volume manipulations. Only random errors are subject to mathematical evaluation because they are statistically reproducible.

The degree of reproducibility is measured by the precision, which is an index of reliability of results. The calculation of precision is based on the Gauss probability curve, which assumes, for repeated runs, that

1. Smaller errors are more frequent than larger errors.
2. The arithmetic mean of the errors is the best value of the correct result.
3. Overall precision is better if the individual divergences are close to the mean than if they are further away.

This third assumption means simply that large divergences, even though they may compensate in the determination of the mean, should be penalized over small divergences. The formula for calculation of the error takes account of this penalty.

The precision of a series of measurements (really the magnitude of the error) is expressed by the standard deviation, whose symbol is sigma (σ). It can be considered a figure of merit of the analytical procedure. The standard deviation is determined in the following manner (notation follows ASTM suggested practices (211)).

A sample solution, in which the concentration of the analyte may or may not be known, is run repeatedly through the procedure (the ASTM recommends at least 16 measurements). The sum of these results is $\Sigma\ x$ and their mean is $\Sigma\ x/n$, where n denotes the number of measurements. The individual deviations from this mean are then determined, without regard to sign, each is squared, and the sum of the squares $\Sigma\ d^2$, is taken. The formula for the standard deviation is then

$$\sigma = \sqrt{\frac{\Sigma\ d^2}{n-1}}$$

When precision results are to be compared to samples of different concentration levels, the standard deviation is expressed as a percentage of the mean. This number, previously called the coefficient of variation, is now referred to as the relative standard deviation (RSD), a term recommended by IUPAC, and is the form generally used in literature descriptions of AAS methods.

The standard deviation σ has several very useful properties. It can be shown that 68% of all repetitive measurements in a series will probably fall within 1σ of the mean, 95% will fall within 2σ of the mean, and 99.7% will fall within 3σ of the mean.

The symbol n, standing for the universe of measurements, needs some comment. To be statistically valid, n must be large enough that its RSD does not differ materially from the RSD of an infinite series of measurements. The number n is generally a subjective choice, influenced by the extent of scatter of the results; a wide scatter indicates the need for more measurements. An objective test of sorts is to divide the series of measurements into two groups, determine the RSD for each, and then judge from the spread whether more measurements are needed.

A second index of precision that is often used is the probable error P of a single determination, the amount on either side of the determined value within which the actual value is as likely as not to fall. The probable error is related to the standard deviation by

$$P = 0.67\sigma$$

That is, if we let x denote the determined value, the actual value has a 50% probability of being in the range $x \pm 0.67\sigma$. Derivations of these expressions cannot be given here; the reader is referred to standard texts on statistics and probability (212–215).

To illustrate the use of these indices of precision, we take as an example a common problem of atomic absorption—the establishment of a calibration curve. Assume that there are samples of four standard concentrations run in triplicate, twelve determinations in all, and that the curve is expected to be linear and pass through the origin. For a measure of repeatability, we cannot use the absorbance readings themselves since they correspond to four different concentrations. However, since we assume that all readings lie on or near the same linear absorbance-concentration curve we can use the slope of the line defined by each reading and the origin. This slope is obtained by dividing the observed absorbance, the dependant variable, by the concentration of the sample.

The results are shown in Table 4.1, where the columns list the run number, concentration, observed absorbance, slope, deviation d of the slope from the mean slope, and d^2. The concentration of an unknown can now be calculated as absorbance/(0.0547 ± 0.00074).

What to do with the outliers—those vexing, exceptional results that spoil an otherwise good precision report. Always present is the temptation to drop these outliers on the ground that they are a nonrecurring accident. Statisticians have not come up with an entirely objective rule; one sometimes reads the suggestion that errors larger than 3σ be dropped on the justification that their probabilities of occurrence (only 0.3%) are so small. Perhaps the better procedure, and certainly the more honest one, is to include the outlying results, or repeat the test series with many more trials and include all the data in the precision figure. The principle underlying the precision assumes that in n trials the result will not change significantly from an infinite number of trials.

For a simple problem such as our example, actual plotting of the curve is unnecessary. Once the slope has been established, unknown concentrations can be read off directly, using a slide rule (if any still exist) or a hand calculator.

Absorbance curves, even though linear for the lower concentrations, invariably curve downward, toward the absorbance axis, at higher concentrations. When it is suspected that the calibration curve is not linear, more and closer spaced points will have to be determined, and the curve drawn freehand. Some AAS instruments have the capability of rectifying

TABLE 4.1 Determination of Slope of Calibration Curve by Means of Standard Deviation

No.	Concentration (μg/ml)	Absorbance	Slope	d	d^2
1	2	0.110	.0550	3×10^{-4}	9×10^{-8}
2		0.105	.0525	22	484
3		0.110	.0550	3	9
4	3	0.164	.0547	1	1
5		0.166	.0553	6	36
6		0.166	.0553	6	36
7	5	0.270	.0540	7	49
8		0.270	.0540	7	49
9		0.285	.0570	23	529
10	8	0.430	.0537	10	100
11		0.440	.0550	3	9
12		0.440	.0550	3	9
			.6565		1320×10^{-8}

Mean slope = .6565/12 = .0547

$$\sigma = \sqrt{\frac{\Sigma d^2}{n-1}} = \sqrt{\frac{1320 \times 10^{-8}}{11}} = 0.0011$$

$$RSD = \frac{.0011}{.0547} \times 100 = 2.0\%$$

$$P = 0.67 \times 0.0011 = 0.00074$$

$$Slope = 0.0547 \pm 0.00074$$

calibration curves electronically after two or three points have been determined on the curved portion and the results fed into the device.

Extrapolations are always uncertain and should not be attempted unless the distance extrapolated is short.

Another standard method of curve fitting, precise but laborious and applicable only to linear functions, is the least-squares treatment. The least-squares operation depends on evaluating the constants a and b in the equation $y = ax + b$ such that divergences from these two parameters are minimums. For curves that pass through the origin, $b = 0$ and only a need be evaluated. Worthing and Geffner (216) describe a procedure, slightly simplified for the calculation, and Rainsford (217) presents a still simpler procedure but one less precise.

Every eighth grader, preparing for his or her first Science Fair project is instructed never, never to change more than one variable at a time. Sherman (218) showed that by the method of covariance analysis, contributions of each of several variables being changed at the same time can still be evaluated. His monograph on statistical analysis, written specifically for spectroanalytical problems, is an excellent treatment of the subject.

4.5.3 PHOTOMETRIC ERROR FUNCTION

Noise in the photometric measuring signal, arising in the receiver-amplifier, the flame, and the sample transport into the nebulizer-burner, causes an error in the photometric reading. The magnitude of the error depends strongly on the point of the scale at which the reading falls, and has nothing to do with the random errors of the analysis.

The first to elaborate on this idea appear to have been Twyman and Lothian (219), followed by Gridgeman (220), who presented a thorough discussion of the concept. Gridgeman's argument goes as follows: The absorbance A by definition is

$$A = - \log T$$

where T is the transmittance. The error is then ΔA and the relative error is $\Delta A / A$. If A is differentiated with respect to T, we obtain

$$\frac{d \, A/A}{d \, T} = - \log \frac{I}{A} = \frac{1}{T \ln T}$$

This is merely an identity based on the original definition. If now we assign an actual value to ΔA and plot the relative error $\Delta A / A$ against T, we obtain the curve of Fig. 4.8, in which $\Delta A = 1\%$ (assumed). The derivative with respect to T yields a minimum at 0.43 absorbance unit, or at about 37% T.

To reduce the reading error, several options are available to the operator. A digital readout will increase the least counts by about a factor of ten times the simple zero-to-100% scale of a galvanometer or strip-chart recorder. Scale expansion will help, but noise is also expanded. Perhaps the most effective option is the integrating readout, which will smooth out variations. Bracketing the unknown closely between standards is also an effective approach.

Skagerboe (221) reports on the results of three techniques of repetitive

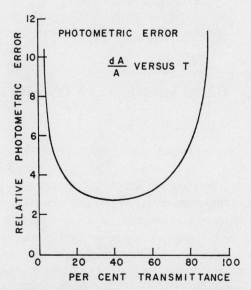

Fig. 4.8 Curve of photometric error versus scale reading. Assumed scale error for the plot has been 1%. For any other value, the actual error at any absorption may be found by multiplying the setting error by the indication taken from the plot. For example, if it is possible to read absorption to 0.5%, the photometric error at 60% will be 1.5%.

measurements of sample concentrations (at various absorbances). He measured peak height with a recorder, peak area by scanning, and counting on peak height by holding the scan on the line. The third technique gave by far the best precision, and followed the $\sqrt{n}$ rule, where n is the number of counts. Counters, however, are not common readouts in AAS equipment.

To test the Gridgeman theory Slavin (222) measured the photometric error of a large number of instruments and found good agreement but with a slightly larger standard deviation at higher absorbances than expected, which he ascribed to mechanisms other than readout errors.

As indicated in Fig. 4.7, the optimum working range is between 20 and 60% transmittance, or about 0.1 to 0.7 absorbance.

CHAPTER

5

PHYSICAL AND CHEMICAL DATA ON THE ELEMENTS

This chapter lists the physical and chemical properties of the individual elements that are of interest in AAS. The physical data include atomic weights, wavelengths of the most sensitive (best) line and secondary lines, type of scanning lamp appropriate to the determination, sensitivity and linear range of the recommended line. The chemical data consist of the weight and identity of the compound to be used to make up the standard solution at a concentration both convenient and suitable for storage.

Most of these data are taken from *Analytical Methods for Atomic Absorption Spectrophotometry*, published by Perkin-Elmer Corporation in 1973 and irreverently known as the Cookbook. Included here is a concise table (Table 5.1) of the elements to which AAS can be applied, the approximate slit width, and the flame gas mixture recommended for the determination. In addition the table contains the concentration for each element, in $\mu g/ml$, which will result in a reading of approximately 0.2 absorbance unit. Although this is not the usual way of presenting this datum, which is generally shown in $\mu g/ml$ to give an absorption of 1% (0.0044 absorbance), the higher value may prove to be more convenient in practice. At any rate the two parameters are easily converted from one to the other.

References to work done on the determination of the various elements might have been a valuable addition to this chapter, but the literature is too voluminous. A good up-to-date source for typical procedures for the elements is another publication of Perkin-Elmer, the Bibliography issued twice annually as part the *Atomic Absorption Newsletter*. It categorizes papers by element, and the full paper titles are included, so that search of the literature can be rapid and convenient.

Notes

All wavelengths are given in nanometers.
The following abbreviations are used throughout:

110

HCL—hollow-cathode lamp
EDL—electrodeless discharge lamp
Ac—acetylene gas
O—oxidizing flame
R—reducing flame

Windows of hollow-cathode lamps, where the principal line wavelength is in the visible, may be of glass. If these lamps are to be used in the ultraviolet for minor lines, window material should be of fused silica.

TABLE 5.1. Standard Conditions for Atomic Absorption[a]

Element	λ (nm)	SBW (nm)	Flame Gases	Sensitivity Check[b]
Ag	328.1	0.7	A–Ac	3
Al	309.3	0.7	N–Ac	45[c]
As	193.7	0.7	A–Ac	40
As	193.7	0.7	N–Ac	45
As	193.7	0.7	Ar–H	7
Au	242.8	0.7	A–Ac	12
B	249.7	0.7	N–Ac	2000
Ba	553.6	0.2	N–Ac	20[c]
Ba	553.6	0.2	A–Ac	225[c]
Be	234.9	0.7	N–Ac	1.2
Bi	223.1	0.2	A–Ac	18
Ca	422.7	0.7	A–Ac	3.5
Ca	422.7	0.7	N–Ac	2.5[c]
Cd	228.8	0.7	A–Ac	1.2
Co	240.7	0.2	A–Ac	7
Cr	357.9	0.7	A–Ac	4
Cs	852.1	2.0	A–Ac	15[c]
Cu	324.7	0.7	A–Ac	4
Dy	421.2	0.2	N–Ac	40[c]
Er	400.8	0.2	N–Ac	45[c]
Eu	459.4	0.2	N–Ac	25[c]
Fe	248.3	0.2	A–Ac	6
Ga	287.4	0.7	A–Ac	120
Nd	463.4	0.2	N–Ac	500[c]
Ni	232.0	0.2	A–Ac	7
Os	290.9	0.2	N–Ac	45
P	213.6	0.2	N–Ac	15000
Pb	283.3	0.7	A–Ac	25
Pd	247.6	0.2	A–Ac	12
Pr	495.1	0.2	N–Ac	2500[c]
Pt	265.9	0.7	A–Ac	90
Rb	780.0	2.0	A–Ac	5[c]
Re	346.0	0.2	N–Ac	700
Rh	343.5	0.2	A–Ac	15
Ru	349.9	0.2	A–Ac	25
Sb	217.6	0.2	A–Ac	25
Sc	391.2	0.2	N–Ac	20[c]
Se	196.0	2.0	A–Ac	25
Se	196.0	2.0	Ar–H	12
Si	251.6	0.2	N–Ac	85
Sm	429.7	0.2	N–Ac	400[c]
Sn	224.6	0.2	A–Ac	110
Sn	224.6	0.2	N–Ac	100
Sn	224.6	0.2	A–H	40
Sr	460.7	0.2	A–Ac	6[c]
Sr	460.7	0.7	A–Ac	4[c]

Gd	407.9	0.2	N–Ac	800[c]
Ge	265.1	0.2	N–Ac	110
Hf	286.6	0.2	N–Ac	700[d]
Hg	253.6	0.7	A–Ac	350
Ho	410.4	0.2	N–Ac	50[c]
In	303.9	0.7	A–Ac	35
Ir	264.0	0.2	A–Ac	400
K	766.5	2.0	A–Ac	1.8
La	550.1	0.2	N–Ac	2000[c]
Li	670.8	0.7	A–Ac	1.5
Lu	336.0	0.2	N–Ac	300[c]
Mg	285.2	0.7	A–Ac	0.3[c]
Mn	279.5	0.2	A–Ac	2.5
Mo	313.3	0.7	N–Ac	25
Mo	313.3	0.7	A–Ac	40
Na	589.0	0.2	A–Ac	0.7
Nb	334.4	0.2	N–Ac	1700[c]

Ta	271.5	0.2	N–Ac	725
Tb	432.6	0.2	N–Ac	400[c]
Tc	261.5	0.2	A–Ac	120
Te	214.3	0.2	A–Ac	45
Ti	365.3	0.2	N–Ac	85
Tl	276.8	0.7	A–Ac	25
Tm	371.8	0.7	N–Ac	16[c]
U	358.5	0.2	N–Ac	2300[c]
V	318.4	0.7	N–Ac	75
W	255.1	0.2	N–Ac	500
Y	410.2	0.2	N–Ac	85[c]
Yb	398.8	0.2	N–Ac	5[c]
Zn	213.9	0.7	A–Ac	0.8
Zr	360.1	0.2	N–Ac	475

[a] SBW = slit bandwidth. Flame gases are:

   A–Ac = Air–acetylene

   N–Ac = Nitrous oxide–acetylene

   Ar–H = Argon–hydrogen–entrained air

   A–H = Air–hydrogen

[b] Metal concentration (μg/ml) in aqueous solution that will give a reading of *approximately* 0.2 absorbance unit with either the 4-in. single-slot air–acetylene burner head or the 2-in. nitrous oxide–acetylene burner head.

[c] Addition of a large amount of an easily ionizable material (e.g., 1000 μg/ml potassium as the chloride) may be required to control ionization.

[d] Fluoride added to obtain the sensitivity check value shown.

113

5.1   ALKALIES: LITHIUM, SODIUM, POTASSIUM, RUBIDIUM, CESIUM

	Atomic Weight	Best Line	Gases	Lamp	Sensi-tivity	Linear Range to
Li	6.940	670.8	Air–Ac O	HCL	0.035	2.0
Na	22.99	589.0	Air–Ac O	HCL	0.015	1.0
K	39.10	766.5	Air–Ac O	HCL	0.04	2.0
Rb	85.48	780.0	Air–Ac O	EDL	0.1	5.0
Cs	132.9	852.1	Air–Ac O		0.3	15.

Secondary Lines

Li	Na	K	Rb	Cs
323.3	330.2	404.4	420.2	455.5
	330.3	404.7	421.6	

Standard Solutions for a Concentration of 1000 μg/ml

Lithium.   Dissolve 5.324 g of lithium carbonate in a minimum of 1:1 HCl and dilute to 1 liter with water.

Sodium.   Dissolve 2.542 g of sodium chloride in 1 liter of water.

Potassium.   Dissolve 1.907 g of potassium chloride in 1 liter of water.

Rubidium.   Dissolve 1.415 g of rubidium chloride in 1 liter of water.

Cesium.   Dissolve 1.267 g of cesium chloride in 1 liter of water.

Notes

Alkali salts are seldom pure. Lithium occurs in small amounts in sodium salts, and rubidium and cesium in potassium salts, even in reagent grade chemicals. Blanks should always be run.

These easily ionizable atoms should have ionization suppressed by the addition of another alkali salt.

5.2 ALKALINE EARTHS: BERYLLIUM, CALCIUM, MAGNESIUM, STRONTIUM, BARIUM

	Atomic Weight	Best Line	Gases	Lamp	Sensi- tivity	Linear Range to
Be	9.013	243.9	N_2O–Ac R	HCL	0.025	4
Ca	40.08	422.7	Air–Ac R	HCL	0.08	7
Mg	24.32	285.2	Air–Ac O	HCL	0.007	0.5
Sr	87.63	460.7	Air–Ac R	HCL	0.12	5
Ba	137.4	553.6	N_2O–Ac R	HCL	0.4	25

Secondary Lines

Be	Ca	Mg	Sr	Ba
—[a]	239.9	202.6	242.8	350.1
		280.0	256.9	455.4[b]
			293.2	553.6

[a] No good lines.
[b] An ion line.

Standard Solutions for a Concentration of 1000 μg/ml

Beryllium. Dissolve 1.000 g of beryllium metal in a minimum volume of 1:1 HCl. Dilute to 1 liter with 1% HCl solution.

Calcium. To 2.498 g of calcium carbonate, $CaCO_3$, add approximately 100 ml of water, then add carefully about 20 ml of concentrated HCl to complete solution. Dilute to 1 liter with water.

Magnesium. Dissolve 1 g of magnesium ribbon in a minimum volume of 1:1 HCl and dilute to 1 liter with 1% HCl.

Strontium. Dissolve 2.415 g of strontium nitrate, $Sr(NO_3)_2$, in 1 liter of 1% HNO_3.

Barium. Dissolve 1.779 g of barium chloride, $BaCl_2 \cdot 2H_2O$, in 1 liter of water.

Notes

Beryllium is extremely sensitive in the furnace, down to about 10^{-13} g.

The alkaline earths tend to emit molecular bands in the cooler flames, which may cause spectral interference if lamp modulation is not used.

5.3 COPPER, SILVER, GOLD

	Atomic Weight	Best Line	Gases	Lamp	Sensitivity	Linear Range to
Cu	63.54	324.7	Air–Ac O	HCL	0.09	5
Ag	107.88	328.1	Air–Ac O	HCL	0.06	4
Au	197.0	242.8	Air–Ac O	HCL	0.25	20

Secondary Lines

Cu	Ag	Au
327.4	338.3	267.6
216.5		312.3
222.6		274.8
217.9		
249.2		

Standard Solutions for a Concentration of 1000 μg/ml

Copper.   Dissolve 1.000 g of copper metal in a minimum volume of HNO_3 and dilute to 1 liter with 1% HNO_3.

Silver.   Dissolve 1.574 g of silver nitrate, $AgNO_3$, in water and dilute to 1 liter with 1% HNO_3. Store in an amber glass bottle.

Gold.   Dissolve 0.1000 g of gold metal in a minimum volume of aqua regia. Evaporate to dryness, redissolve in a minimum volume of HCl, and dilute to 100 ml with water. Store in an amber glass bottle.

5.4 TRANSITION METALS: VANADIUM, CHROMIUM, MANGANESE, IRON, COBALT, NICKEL

	Atomic Weight	Best Line	Gases	Lamp	Sensi- tivity	Linear Range to
V	50.95	318.3^a	Air–Ac R	HCL	1.7	150
Cr	52.01	357.9	Air–Ac R	HCL	0.1	5
Mn	54.94	279.5	Air–Ac O	HCL	0.055	3
Fe	55.85	248.3	Air–Ac O	HCL	0.12	5
Co	58.94	240.7	Air–Ac O	HCL	0.15	5
Ni	58.71	232.0	Air–Ac O	HCL	0.15	5

^{a}This is a triplet: 318.3, 318.4, 318.5 mm.

Secondary Lines

V	Cr	Mn	Fe	Co	Ni
306.6	359.4	279.8	248.8	242.5	231.1
306.0	360.5	280.1	302.1	241.2	252.5
305.6	425.4	403.1	302.2	252.1	341.5
320.2	427.5	321.7	252.7	243.6	305.1
390.2	429.0		372.0	304.4	346.2
			373.7	352.7	351.5
			344.1	346.6	303.8
			305.9	347.4	337.0
			346.6	301.8	323.3
			392.0		294.4

Standard Solutions for a Concentration of 1000 μg/ml

Vanadium. Dissolve 1.000 g of vanadium metal in a minimum volume of HNO_3 and dilute to 1 liter with 1% HNO_3.

Chromium. Dissolve 3.735 g of potassium chromate, K_2CrO_4, in 1 liter of water.

Manganese. Dissolve 1.000 g of manganese metal in a minimum volume of HNO_3 and dilute to 1 liter with 1% HCl.

Iron. Dissolve 1.000 g of iron wire in 1:1 HNO_3 and dilute to 1 liter with water.

Cobalt. Dissolve 1.000 g of cobalt metal in a minimum volume of 1:1 HCl and dilute to 1 liter with 1% HCl.

Nickel. Dissolve 1.000 g of nickel metal in a minimum volume of 1:1 HNO_3 and dilute to 1 liter with 1% HNO_3.

5.5   LOW-MELTING METALS: ZINC, GALLIUM, GERMANIUM, CADMIUM, INDIUM, TIN, MERCURY, THALLIUM, LEAD

	Atomic Weight	Best Line	Gases	Lamp	Sensi- tivity	Linear Range to
Zn	65.38	213.9	Air–Ac O	HCL	0.018	1
Ga	69.72	287.4	Air–Ac O	HCL	2.5	200
Ge	72.60	265.1	N_2O–Ac R	HCL	2.4	250
Cd	112.4	228.8	Air–Ac O	HCL	0.025	2
In	114.8	303.9	Air–Ac O	HCL	0.7	50
Sn	118.7	224.6	Air–Ac R	HCL	2.4	200
Hg	200.6	253.6	Air–Ac O	HCL	7.5	300
Tl	204.4	276.8	Air–Ac O	HCL	0.5	20
Pb	207.2	283.3	Air–Ac O	HCL	0.5	20

Secondary Lines

Zn	Ga	Ge	Cd	In	Sn	Hg	Tl	Pb
307.6	294.4	259.2	326.1	325.6	286.3	185.0	377.6	217.0
	417.2	271.0		410.5	235.5		238.0	261.4
	250.0	275.5		451.1	270.6		258.0	368.4
	254.0	269.1		256.0	303.4			
	245.0			275.4	254.7			
					219.9			
					300.9			
					233.5			

Standard Solutions for a Concentration of 1000 μg/ml

Zinc.   Dissolve 1.000 g of zinc metal in a minimum volume of 1:1 HCl and dilute to 1 liter with 1% HCl.

Gallium.   Dissolve 1.000 g of gallium metal in a minimum volume of aqua regia and dilute to 1 liter with 1% HCl.

Germanium.   To 0.1000 g of germanium metal in a Teflon beaker, add 5 ml conc. HF. Add HNO_3 dropwise to complete dissolution and dilute to 100 ml with water. Store in a polyethylene bottle.

Cadmium.   Dissolve 1.000 g of cadmium metal in 1 : 1 HCl and dilute to 1 liter with 1% HCl.

Indium.   Dissolve 1.000 g of indium metal in 1 : 1 HCl and dilute to 1 liter with 1% HCl.

Tin.   Dissolve 1.000 g of tin metal in 100 ml of concentrated HCL and dilute to 1 liter with water.

Mercury.   Dissolve 1.080 g of mercuric oxide, HgO, in 1 : 1 HCl and dilute to 1 liter with water.

Thallium.   Dissolve 1.303 g of thallium nitrate, $Tl(NO_3)$, in 1 liter of water.

Lead.   Dissolve 1.598 g of lead nitrate, $Pb(NO_3)_2$, in 1 liter of water.

5.6   TITANIUM, ZIRCONIUM, HAFNIUM

	Atomic Weight	Best Line	Gases	Lamp	Sensi-tivity	Linear Range to
Ti	47.90	365.4	N₂O–Ac R	HCL	1.9	200
Zr	91.22	360.1	N₂O–Ac R	HCL	10.	800
Hf	178.5	286.6	N₂O–Ac R	HCL	15.	500

Secondary Lines

Ti	Zr	Hf
364.3	354.8	307.3
320.0	303.0	289.8
363.6	301.2	296.5
335.5	298.5	
375.3	362.4	
334.2		
399.9		

Standard Solutions

Titanium.   For a concentration of 1000 μg/ml, dissolve 1.000 g of titanium metal in 100 ml of 1:1 HCl; cool and dilute to 1 liter with water. For further dilution, use 10% HCl.

Zirconium.   For a concentration of 10,000 μg/ml, place 1.000 g of zirconium metal in a Teflon beaker, add 10 ml of water, and then add HF dropwise to complete solution. Dilute to 100 ml with 2% HF. Store in a polyethylene bottle.

Hafnium.   For a concentration of 10,000 μg/ml, use the same procedure as for zirconium to put 1.000 g of hafnium metal in solution and dilute to 100 ml.

5.7 BORON, ALUMINUM, SILICON

	Atomic Weight	Best Line	Gases	Lamp	Sensitivity	Linear Range to
B	10.82	249.7	N_2O–Ac R	HCL	40	200
Al	26.98	309.3	N_2O–Ac R	HCL	1.0	50
Si	28.09	251.6	N_2O–Ac R	HCL	1.8	150

Secondary Lines

B	Al	Si
—[a]	396.2	250.7
	308.2	252.8
	394.4	252.4
	237.31⎤	221.7
	237.34⎦	221.1
	237.6	
	257.5	

[a]No good lines.

Standard Solutions

Boron. For a concentration of 5000 μg/ml, dissolve 28.60 g of boric acid, H_3BO_3, in 1 liter of water and store in a polyethylene bottle.

Aluminum. For a concentration of 1000 μg/ml, dissolve 1.000 g of aluminum wire in a minimum volume of 1:1 HCl, adding a small drop of mercury to the solution as a catalyst. Decant to remove the mercury, and then dilute to 1 liter with 1% HCl.

Silicon. For a concentration of 1000 μg/ml, dissolve 5.056 g of sodium metasilicate, $Na_2SiO_3 \cdot 9H_2O$, in 300 ml of water. Add enough HCl to lower the pH to about 5 and dilute to 1 liter with water. Store in a polyethylene bottle.

5.8 ARSENIC, ANTIMONY, BISMUTH

	Atomic Weight	Best Line	Gases	Lamp	Sensi- tivity	Linear Range to
As	74.91	193.7	Air–Ac O	EDL	0.8	50
Sb	121.76	217.6	Air–Ac O	HCL	0.5	40
Bi	209.0	223.1	Air–Ac O	HCL	0.4	50

Secondary Lines

As	Sb	Bi
189.0	206.8	222.8
197.2	231.2	306.8
		206.2
		227.7

Standard Solutions for a Concentration of 1000 μg/ml

Arsenic. Dissolve 1.320 g of arsenious oxide, As_2O_3, in 25 ml of 20% (w/v) KOH. Neutralize to a phenolphthalein endpoint with 20% H_2SO_4, and dilute to 1 liter with 1% H_2SO_4.

Antimony. Dissolve 2.743 g of potassium antimony tartrate hemi-hydrate, $K(SbO)C_4H_4O_6 \cdot \frac{1}{2}H_2O$, in 1 liter of water.

Bismuth. Dissolve 1.000 g of bismuth metal in a minimum volume of 1:1 HNO_3 and dilute to 1 liter with 2% HNO_3.

5.9 MOLYBDENUM, TECHNETIUM, RHENIUM

	Atomic Weight	Best Line	Gases	Lamp	Sensi-tivity	Linear Range to
Mo	95.95	313.3	N_2O–Ac R	HCL	0.5	60
Tc	99.	261.4 } 261.6 }	Air–Ac R	HCL	3.	60
Re	186.2	346.0	N_2O–Ac R	HCL	15.	1000

Secondary Lines

Mo	Tc	Re
317.0	260.9	346.5
379.8	429.7	345.2
319.4	426.2	
386.4	318.2	
390.3	423.8	
315.8	363.6	
320.9	317.3	

Standard Solutions for a Concentration of 1000 μg/ml

Molybdenum. Dissolve 1.840 g of ammonium paramolybdate, $(NH_4)_6$ $Mo_7O_{24} \cdot 4H_2O$, in 1 liter of 1% NH_4OH.

Technetium. Stock solution of Tc obtainable from Nuclear Sciences Div., International Chemical and Nuclear Corp., Pittsburgh, Pa.

Rhenium. Dissolve 1.554 g of potassium perrhenate, $KReO_4$, in 200 ml of water and dilute to 1 liter with 1% H_2SO_4.

Note

A method for the determination of technetium has recently been published by Hareland, Ebersole, and Ramachandran (223).

5.10   PLATINUM GROUP: RUTHENIUM, RHODIUM, PALLADIUM, OSMIUM, IRIDIUM, PLATINUM

	Atomic Weight	Best Line	Gases	Lamp	Sensi- tivity	Linear Range to
Ru	101.1	349.9	Air–Ac O	HCL	0.5	50
Rh	102.9	343.5	Air–Ac O	HCL	0.3	50
Pd	106.4	247.6	Air–Ac O	HCL	0.25	15
Os	190.2	290.9	N_2O–Ac R	HCL	1.	200
Ir	192.2	264.0	Air–Ac R	HCL	8.	1000
Pt	195.1	265.9	Air–Ac O	HCL	2.	75

Secondary Lines

Ru	Rh	Pd	Os	Ir	Pt
372.8	369.2	244.8	305.9	208.9	306.5
379.9	339.7	276.3	263.7	266.5	283.0
392.6	350.2	340.5	301.8	237.3	293.0
	365.8		330.2	285.0	273.4
	370.1		271.5	250.3	270.2
	350.7		280.7	254.4	248.7
			264.4	351.4	299.8
			442.0		271.9

Standard Solutions

Ruthenium.   For a concentration of 1000 μg/ml, dissolve 0.2052 g of ruthenium chloride, $RuCl_3$, in a minimum volume of 20% HCl and dilute to 100 ml with 20% HCL.

Rhodium.   For a concentration of 1000 μg/ml, dissolve 0.386 g of ammonium hexachlororhodate, $(NH_4)_3$ $RhCl_6 \cdot 1\frac{1}{2}$ H_2O, in a minimum volume of 10% HCl and dilute to 100 ml with 10% HCl.

Palladium.   For a concentration of 1000 μg/ml, dissolve 0.100 g of palladium wire in a minimum volume of aqua regia and carefully evaporate to dryness. Take up in 5 ml conc. HCl plus 25 ml of water. Heat to dissolution and dilute to 100 ml with water.

Osmium. This solution is obtainable at a $1.01M$ concentration (1902 μg/ml) from the G. Frederick Smith Co., Columbus, Ohio. Solutions must be stored in glass bottles. Further dilutions to be made with 1% H_2SO_4. Osmium solutions are extremely toxic and volatile.

Iridium. For a concentration of 5000 μg/ml, dissolve 1.147 g of ammonium chloroiridate, $(NH_4)_2$ $IrCl_6$, in a minimum volume of 1% HCl and dilute to 100 ml with 1% HCl.

Platinum. For a concentration of 1000 μg/ml, dissolve 0.1000 g of platinum metal in a minimum volume of aqua regia and carefully evaporate to dryness. Take up in 5 ml conc. HCl and 0.1 g NaCl, repeat the evaporation to dryness, and dissolve the residue in 20 ml 1:1 HCl. Dilute to 100 ml with water.

5.11 HEAVY METALS: TUNGSTEN, THORIUM, URANIUM

	Atomic Weight	Best Line	Gases	Lamp	Sensitivity	Linear Range to
W	183.9	255.1	N_2O–Ac R	HCL	11	1500
Th	232.1[a]					
U	238.1	358.5	N_2O–Ac R	HCL	50	6000

[a] Kirkbright, Rao, and West discuss an indirect method for thorium (224).

Secondary Lines

W	Th[a]	U
294.4		356.7
268.1		351.5
272.4		
294.7		
400.9		
283.1		

[a] No data.

Standard Solutions for a Concentration of 10,000 μg/ml

Tungsten. Dissolve 17.95 g of sodium tungstate, $Na_2\,WO_4 \cdot 2H_2O$, in 200 ml of water. Add 100 ml of 10% NaOH solution and dilute to 1 liter with water. Store in a polyethylene bottle.

Thorium. Dissolve 2.380 g of thorium nitrate, $Th(NO_3)_4 \cdot 4H_2O$, in 100 ml of water.

Uranium. Dissolve 21.10 g of uranyl nitrate, $UO_2(NO_3)_2 \cdot 6H_2O$, in 1 liter of water.

5.12   PHOSPHORUS, SELENIUM, TELLURIUM

	Atomic Weight	Best Line	Gases	Lamp	Sensitivity	Linear Range to
P	30.98	213.6	N_2O–Ac R	HCL	260	10,000
Se	78.96	196.0	Air–Ac O	EDL	0.5	50
Te	127.6	214.3	Air–Ac O	HCL	1.0	25

Secondary Lines

P	Se	Te
214.9	204.0	225.9
	206.3	238.6
	207.5	

Standard Solutions

Phosphorus.   For a concentration of 50,000 μg/ml, dissolve 21.32 g of dibasic ammonium phosphate, $(NH_4)_2HPO_4$, in 100 ml of water.

Selenium.   For a concentration of 1000 μg/ml, dissolve 1.000 g of elemental selenium in a minimum volume of concentrated nitric acid. Take to dryness, add 2 ml of water, and evaporate to dryness again. Repeat two or three times and then dissolve in 10% HCl and dilute to 1 liter with water.

Tellurium.   For a concentration of 1000 μg/ml, dissolve 1.000 g of elemental tellurium in a minimum volume of concentrated nitric acid. Dilute with 50 ml of water, redissolve the precipitate with a minimum of concentrated hydrochloric acid, heat to drive off the nitrogen oxides, and then dilute to 1 liter with 1% HCl.

5.13   RARE EARTHS

SCANDIUM	SAMARIUM	HOLMIUM
YTTRIUM	EUROPIUM	ERBIUM
LANTHANUM	GADOLINIUM	THULIUM
PRASEODYMIUM	TERBIUM	YTTERBIUM
NEODYMIUM	DYSPROSIUM	LUTETIUM

	Atomic Weight	Best Line	Sensitivity (μg/ml-%)	Linear Range	Standards		
					Weight of Oxide	Volume	Concentration (μg/ml)
Sc	44.96	391.2	0.4	25	3.067	1000	2000
Y	88.92	410.2	1.8	200	1.270	1000	1000
La	138.9	550.1	45.	2500	1.173	1000	1000
Pr	140.9	495.1	55.	400	1.170	100	10,000
Nd	144.3	463.4	10.	1000	1.167	100	10,000
Sm	150.4	429.7	8.5	500	1.159	100	10,000
Eu	152.0	459.4	0.5	50	1.158	1000	1000
Gd	157.3	407.9	16.	1000	1.153	100	10,000
Tb	158.9	432.6	9.	400	1.176[a]	100	10,000
Dy	162.5	421.2	0.9	20	1.148	1000	1000
Ho	164.9	410.4	1.	100	1.146	1000	1000
Er	167.3	400.8	1.	40	1.143	1000	1000
Tm	168.9	371.8	0.35	60	1.142	1000	1000
Yb	173.0	398.8	0.1	20	1.139	1000	1000
Lu	175.0	336.0	6.	500	1.137	100	10,000

[a]The terbium oxide is Tb_4O_7; all the others are the sesquioxide (R_2O_3).

Secondary Lines

Sc	Y	La	Pr	Nd	Sm	Eu	Gd
390.8	407.7	418.7	513.3	492.4	476.0	462.7	368.4
402.4	412.8	495.0					378.3
402.0	414.3	357.4					405.8
405.5	362.1	365.0					405.4
327.0		392.8					371.4
408.2							419.1
327.4							367.4
							404.5

Tb	Dy	Ho	Er	Tm	Yb	Lu
431.9	404.6	405.4	386.3	410.6	346.4	331.2
390.1	418.7	416.3	415.1	374.4	246.5	337.7
406.2	419.5	417.3	389.3	409.4	267.2	356.8
433.8	416.8	404.1	408.8	418.8		298.9
410.5		410.9	393.7	420.4		451.9
		412.7	381.0	375.2		
		422.7	390.5	436.0		
		413.6	394.4	341.0		
		395.6				

Standard Solutions

For the concentration indicated, use the weights shown and make up to volume specified. Except for terbium, all the rare earths are available in the form of the sesquioxide. Dissolve in a small amount of HCl (cautiously) and dilute to specified volume with 1% HCl.

Notes

Cerium has been omitted because there is no good line to use. It can be determined indirectly, by the method of Johnson et al. (225), using the molybdocerophosphate compound.

The rare earths require the addition of an easily ionized element (usually potassium) to suppress ionization. Its concentration should be about 1000 μg/ml.

Rare earth oxides of high purity can be obtained from any of the following suppliers:

Lindsay Rare Earths; Kerr-McGee Chemical Corp., 258 Ann St., West Chicago, IL 60185

Matheson, Coleman and Bell, East Rutherford, N.J. 07073

Molybdenum Corporation of America, 280 Park Ave., New York, NY 10017

Alfa Inorganics, 8 Congress St., Beverly, MA 01915

5.14   ELEMENTS DETERMINABLE BY INDIRECT METHODS

Certain of the elements, which may be insensitive by usual AAS methods, or whose sensitive lines fall in the ultraviolet, outside the usual range and reached only by special equipment, can still be determined satisfactorily by indirect methods. A case that immediately comes to mind is sulfur, which can be precipitated as barium sulfate; either the precipitated barium or the excess remaining in the filtrate can then be measured. The same procedure could be applied to the determination of chlorine, bromine, and iodine, using the silver salt; however, in a mixture of the halides, the precipitate will not be specific owing to the differences in atomic weights.

Another precipitation method, applicable to a group of metals that are not very sensitive in themselves, depends on the formation of a phosphomolybdate complex. After separation, the molybdenum is measured by conventional AAS techniques. This complex is formed by Ta, Nb, Ce, W, Th, U, and several other metals. In general, any stoichiometric compound that can be separated from an excess of reagent can be employed.

Kirkbright and Johnson (226) have reviewed the literature on indirect methods.

APPLICATIONS

This chapter contains abstracts of 42 papers on specific analytical problems that are concerned primarily with the sample type and secondarily with elements. The problems are mainly chemical, dealing with such factors as the matching of matrices or the concentration of the analyte to a level convenient for measurement. Often, these two problems can be solved simultaneously. Once the sample is in an appropriate solution at a concentration suitable for measurement, the measurement procedure is generally conventional.

Papers presented here have been published recently in English (except for two in German) in readily available journals. Since they were published in leading journals, they presumably contain some element of novelty.

The literature of AAS is so voluminous that space was the limiting consideration. Other excellent sources of information on applications are two recently published books: the textbook by Kirkbright and Sargent (227) which has 176 pages on applications, with numerous references, and the third volume of the three-volume set edited by Dean and Rains (228). The latter book made its appearance in late 1976, and is devoted entirely to applications.

Review articles are yet another valuable source of information on applications and their problems. Some recent general reviews are by Rains (229), by West (230) and by Hieftje, Copeland and de Olivares (231). Specialized reviews on biological applications have been published by Willis (232), Dawson (233) and Reinhold (234).

Analysis for toxic materials has been reviewed thoroughly by several authors (235–237), and the treatment of metals and alloys by Price (238). An extensive review of methods for the determination of trace amounts in extremely low concentrations has been published by Törg (239).

The abstracts in the present chapter are arranged in five sections according to the field of interest, as follows:

6.1. Industrial Applications.

6.2. Biological Applications, Including Foods and Animal Feeds.

6.3. Geological and Metallurgical Applications.

6.4. Environmental Applications—Air, Waters, and Food Contamination.

6.5. Miscellaneous Applications.

Abbreviations for certain organic extractants and solvents, mentioned in the abstracts, are for the following compounds:

MIBK—methyl isobutyl ketone

TTA—Theroyl trifluoroacetone

APDC—Ammonium pyrrolidine dithiocarbamate

DDTC—diethyl dithiocarbamate

Determination of Cu, Pb, Cd and Mn in Pulp and Paper by Direct Atomization

F. J. Langmyhr, V. Thomassen, and A. Massoumi, *Anal. Chim. Acta*, **68**, 305 (1974)

Sampling of both pulp and paper was done by punching out small disks with an ordinary office hole-punch. Weight of a single disk was arrived at by weighing ten disks and averaging. Variation in weights was found to be about 4%.

For the analysis, samples of one to ten disks, with a total weight of 1 to 20 mg, were placed in a tantalum boat and inserted into the graphite tube of a furnace. Standard solutions, where necessary, were added to the samples by means of a pipet through the radial hole in the tube. Drying, ashing, and atomizing steps followed.

Precision of the results varied from 8.5% up to 45%. However, the direct method proved to be very quick and simple compared to the conventional wet-ashing preparation. There was little or no danger of contamination, as no reagents were added.

Determination of Total Chromium in Coatings (Paint) by Atomic Absorption Spectroscopy

R. J. Noga, *Anal. Chem.*, **47**, 332 (1975)

The object of this investigation was to find a method of putting paint coatings into solution as an alternative to fusion of the ash in alkali salts, which results in undesirably high concentrations of solids in the solution.

The procedure adopted was to ash the sample in a muffle furnace at 450°C, weigh it to determine percent ash, and then dissolve the ash in a pressure vessel (Paar 4745 Acid Digestion Bomb) with acid. The acid solution consisted of 0.200 g $KMnO_4$ dissolved in 100 ml 1 : 1 H_2SO_4. The sample was digested at 120°C for 1.5 hr; then the excess $KMnO_4$ (which must be in excess) was neutralized with 0.1% sodium azide solution. The Cr was then determined in this solution by the conventional AAS procedure, in a nitrous oxide–acetylene flame.

For comparison with the fusion procedure, a synthetic paint standard was run by both methods. The bomb digestion method gave a precision of 2.5% at the 95% confidence level; the fusion method gave results that were too high by about 25%, owing to interference because of the presence of salts. The detection limit by the bomb method was about 1μg/ml of Cr.

Comments

The presence of fusion salts in the solution would have required that the chromium be separated out by organic extraction or alternatively, that the solution be diluted to reduce interferences by the high concentration of salts.

For the analogous problem of lead in paint, see reference 240.

Determination of Gold in Photographic Film by Flameless Atomic Absorption (in German)

K. Dittrich and W. Mothes, *Talanta,* **22,** 318 (1975)

This paper describes a procedure for the determination of very small amounts of gold that had been added to the photographic emulsion as a sensitizer. As the total gold content in the samples of emulsion was only in the range 10^{-7} to 10^{-8} g, a very sensitive method was indicated. The furnace technique proved adequate.

The gelatine emulsion was decomposed by the action of an enzyme, followed by a treatment with nitric acid plus hydrogen peroxide. From this solution, the gold was extracted into a measured volume of MIBK, which had been saturated with HBr, and then atomized in a tube furnace.

The Au wavelength used was 242.8 nm. The samples consisted of chips of 1 and 2 cm². A weight of gold of 0.17 ng gave an absorption of 1% (0.0044 absorbance unit). The calibration curve was linear to about 10 ng, and the RSD at 5 ng was 11%.

Simultaneous Determination of Trace Wear Metals in Used Lubricating Oils by Atomic Absorption Spectroscopy Using a Silicon-Target Vidicon Tube

R. W. Jackson, K. M. Aldous, and D. G. Mitchell, *Appl. Spectry.*, **28**, 569 (1974)

Sprague and Slavin (241) long ago described a simple method for this determination by conventional AAS, but the various elements had to be measured serially, as no appropriate instrument was available to do the job simultaneously. The serial method is very troublesome, requiring wavelength change, amplifier gain setting, and change of exciting lamp, slit width, and flame height for each element determined. In addition, at least one standard must be run to guard against working curve shift.

For this investigation the advantages of AAS and the emission direct reader were combined. The photomultiplier of the single-channel AAS equipment was replaced by a silicon-target vidicon television camera (242), a comparatively simple change. The resulting vidicon spectrometer gives sensitivities equivalent to the single-channel equipment, although detection limits and dynamic range are somewhat poorer.

Details and description of the vidicon spectrometer are given in the above reference. Performance was tested with standards of Ni, Co, Fe, Mn, and Cu, using a multielement hollow cathode lamp containing these metals. Another multielement lamp served for the measurement of Zn, Cd, Pb, and Ag. Standards for all of these metals as the sulfonates were prepared in clean oil in a concentration range of 0.1 to 100 μg/g.

Resolution of the vidicon, because of the small dispersed image, is only about 1 nm, so spectral interferences are more likely than with single-channel equipment. For FE, Mn, Cu, and Ag, results by vidicon showed an RSD about twice as large as those with conventional AAS. Twenty different real samples run by both techniques showed good agreement.

One problem that may be encountered by multielement analysis is the great difference in concentrations between elements. Thus, a knowledge of the working range for each element at a specific wavelength (which may be subject to spectral interference) is important.

The vidicon technique is not suitable for use with the nitrous oxide–acetylene flame because of the unmodulated signal (the vidicon tube is too slow an integrator), which would cause too much emission interference. A further problem would be detector saturation due to intense band emission of the flame.

Comments

This is a good exposition of the advantages and disadvantages of multielement analysis, both in general and with a vidicon tube. Conventional emission with a direct reading spectrometer still seems more promising, although the sample preparation procedure will have to be changed to end up with a solid.

Since wear indications need be only qualitative, with only two answers (above or below a limiting value), an emission technique with photographic recording and no photometry may possibly be sufficient.

Trace Metals Analysis in Petroleum Products by Atomic Absorption

M. S. Vigler and V. F. Gaylor, *Appl. Spectry.*, **28**, 342 (1974)

The aim of this paper was to investigate the feasibility of preparing a solution of metallic impurities in various petroleum products by a simple ashing technique. The analytes were 23 metals commonly or seldom occurring in petroleum and its distillates. The ashing technique involved the absorption of liquid samples on magnesium sulfonate ash aid (Conostan, Conoash M) or potassium sulfonate, followed by ignition in a muffle at 650°C and solution of the resulting ash in an appropriate mineral acid. The solution was then aspirated and the atomic absorption measured.

The refractory metals required a nitrous oxide–acetylene flame in a one-slot 2 in. burner; the other metals were aspirated into an air–acetylene flame in a 4 in. burner.

Volume of sample taken for the ignition was adjusted to the assumed concentration level of the analyte. This concentration was reduced by the degree of refinement of the petroleum product, with the light distillates requiring a sample of 35 g to assure a measurable quantity of analyte. Ashing time at 650°C varied from 30 min to several hours.

The important factor in this simple ashing process is obviously the degree of retention of the trace metals in the ash. Spiked samples of Al, Sb, Be, Cd, and Pb showed generally good retention, varying from 100% down to 78%, on a 1 ppm standard.

Conoash M, the sulfonate most used, is a product of the Continental Oil Company, developed for analysis of petroleum products by X-ray fluorescence.

Comments

Gorsuch (243) and Boar and Ingram (244) also studied the ashing technique for petroleum samples. General analytical procedures were reported on by Sprague and Slavin (241), Smith et al. (245), and Chuang and Winefordner (246), who also suggested the use of aqueous standards in place of organometallic compounds.

Rapid Determination of Lead in Gasoline by Atomic Absorption Spectroscopy in the Nitrous Oxide–Hydrogen Flame

R. J. Lukasiewicz, P. H. Bernes, and B. E. Buell, *Anal. Chem.*, **47**, 1045 (1975)

Earlier methods for determining lead content in gasoline have the drawback that the response depends on the form of alkyl type to which the lead is bound. Therefore, inorganic lead standards cannot be used because of the different burning characteristics of organic and inorganic standards. According to this paper, the use of nitrous oxide–hydrogen flame allows for the direct aspiration of gasoline samples with no non-atomic absorption at the usually used line at 283.3 nm.

The inorganic standard is made by dissolving reagent grade $PbCl_2$ in a mixture of 10% aliquot solution No. 336 and 90% MIBK. Aliquot 336 is obtainable from General Mills Chemicals, Inc., Minneapolis, Minn.

The standard solution is made up to a strength of 5 g/gallon, with working standards made by dilution with lead-free gasoline. The limit of detection is approximately 0.0001 g/gallon, and the linear range extends to 0.10 g/gallon. This should be the working range.

Samples for analysis were prepared by adding a small quantity of iodine dissolved in toluene to the sample; this was allowed to react for 5 min and then aliquot 336 in MIBK was added. The samples thus prepared were measured by direct comparison with the $PbCl_2$ standards in lead-free gasoline.

The results in precision are comparable to those obtained by the ASTM recommended method [(247, 248)] namely ±4%, average 2%. The nitrous oxide–hydrogen flame is convenient to use with gasoline; in fact, a steady flame can be obtained with the gasoline alone as fuel, with no hydrogen.

Determination of Lead in Gasoline using a Total Cunsumption Burner

L. L. McCorriston and R. K. Ritchie, *Anal. Chem.*, **47**, 1137 (1975)

This method is especially applicable to low lead concentrations. It has the advantage over the ASTM method of not requiring any pretreatment of the sample. It employs a total consumption burner, an isooctane-acetone solvent, and lead-alkyl standards. The flame gases are air–hydrogen.

Commercial gasoline comes in three grades of lead concentration: regular/premium, low-lead, and no-lead. The procedure consisted of diluting all three grades to a concentration in the range of 0.05 g Pb per U.S. gallon, so that the lead concentration would fall on the linear portion of the working curve. With this method, a solution containing 0.05 g/gal has an absorbance of 0.21, and lead concentrations can be read directly, after correction for background with a deuterium lamp. The wavelength used was 283.3 nm.

Precision was reported to be excellent, with standard deviations of 0.002, 0.019, and 0.073 at the 95% confidence level for the three gasoline grades.

Comments

This and the previous paper appeared in the same issue of *Analytical Chemistry*, and it is interesting to compare the solutions of the analytical problem by the two sets of workers. In the paper by Lukasiewicz et al., the object was to find a way of using inorganic standards, although with some pretreatment. McCorriston and Ritchie accept the use of lead alkyls as standards, but then operate with no additional preparation. Note, though, that the burner will have to be replaced with a lamellar flow type if other elements are to be run.

Determination of Trace Elements in Fish Tissues by the Standard Addition Method

Kaare Julshamn and Olaf R. Braekkan, *At. Absorpt. Newsl.*, **14**, 49 (1975)

This was one of a series of papers on trace elements in marine food organisms. The elements of interest were Fe, Zn, Cu, Mn, Pb, and Cd. The problems encountered were mainly in the destruction of the organic matter and the concentration of the analytes.

Preparation consisted in freeze-drying and homogenizing the samples. Samples of 0.25 g were wet-digested by heat and pressure, chelated by NDDC, and extracted into MIBK. These solutions were then atomized conventionally, except for the Pb and Cd, which were measured by the sampling boat technique. The type of flame was not stated.

The lack of standards led to the use of the method of standard additions. Of four replicate volumes of the unknown solution, three were spiked with the analyte metals. All points fell on linear calibration curves and were extrapolated to the abscissa in the usual manner. Recoveries were between 90 and 104%; standard deviations were not reported.

Note

Reference 249 is on the same subject.

Determination of Cobalt and Copper in Animal Feeds by Extraction and Atomic Absorption Spectroscopy*

S. A. Popova, L. Bezur, and E. Pungor, *Z. Anal. Chem.*, **271**, 269 (1974)

Cobalt and copper occur in very low concentrations in some animal feeds; some form of concentration must be employed to bring the metal content to a measurable level. One well-tried method for other types of sample has been the use of APCD (ammonium pyrolidine dithiocarbamate) as a complexing agent and MIBK as an extractant. This process appears not to have been tried on animal feeds.

The feeds treated were corn, barley, and soybean flours and meat meal.

*Published in English.

The flours were dry-ashed at 550°C; the meat meal was partially wet-ashed in HNO_3, dried, and finally ashed in the muffle. The ashes were dissolved in HNO_3, the acid replaced by HCl after two evaporations; then the solution was filtered and the pH of the filtrate adjusted to about 3. This solution was then extracted with the APCD and MIBK mixture. After separation, the organic phase was aspirated into the air–acetylene flame.

Wavelengths used were 240.7 nm for Co and 324.7 nm for Cu. The concentration range examined was 3 to 25 ppm for Co and 150 to 850 ppm for Cu. At these levels the RSD was about 8% for Co and 7% for Cu. The APDC extraction concentration method is advantageous because the pH is not critical.

Comments

No mention was made of the extent of recovery, although other workers have pointed out the dangers in both dry- and wet-ashing (see Section 4.1.2).

Determination of Trace Amounts of Copper and Zinc in Edible Fats and Oils by Acid Extraction and Atomic Absorption Spectroscopy

R. A. Jacob and L. M. Klevay, *Anal. Chem.*, 47, 741 (1975)

Trace metals in fats and oils in concentrations as low as 30 ppb are harmful, causing odor and degradation of color. Direct aspiration of the sample into a flame, after dilution with a solvent does not produce the required sensitivity for measurement. The procedure outlined here involves acid extraction with EDTA concentration, followed by aspiration into a flame and conventional AAS determination.

Recovery studies were made with low metal content soybean oil as a diluent for several National Bureau of Standards organometallic standards. Lines used were 324.7 nm for Cu and 213.9 nm for Zn. The flame was air–acetylene.

Procedure: Fifty grams of sample were refluxed with 300 ml of extractant (18% HCl × 0.01% EDTA); 50 ml then was removed from the lower aqueous layer, and the action repeated with fresh extractant. The two extracts were combined and diluted with water, concentrated HNO_3 was added, and the mixture was boiled down to a volume of 1 to 2 ml. This was then diluted to either 10 or 25 ml and constituted the working solution, representing the 50 g sample. Recoveries by this procedure were 96% for Cu and 93% for Zn. This was slightly better than the recoveries by char-ashing methods, which are painstaking and slow, requiring 3 to 4 days.

The calculated RSD varied from a maximum of 19% for the 30 ppb level down to about 4% for the higher concentrations, for both metals. Detection limits, expressed as twice the standard deviation, were 0.012 μg for Cu and 0.009 μg for Zn. The procedure can be completed in 4 hr, and several samples can be run per day.

Oxygen-Rich Atmosphere for the Direct Determination of Copper in (Edible) Oils by Non-Flame Atomic Absorption Spectroscopy

M. K. Kundu and A. Prevot, *Anal. Chem.*, **46**, 159 (1974)

The presence of copper in edible oils enhances the tendency to rancidity. The low level of Cu in these oils (about 10^{-8} g/g) calls for the furnace technique to provide sufficient sensitivity for its determination. However, when an oil sample is subjected to direct volatilization in a furnace, the heavy smoke produced interferes with the measurement.

This problem was investigated by trying to ash the sample in the graphite tube of the furnace by mixing the nitrogen purge gas with oxygen and controlling temperature. It was found that a mixture of 1:2 nitrogen–oxygen at 490°C completely destroys the organic matter in a treatment for 90 sec. After this, the temperature is raised and the analysis proceeds in the conventional manner.

A question that must be raised for this technique is the damage caused to the furnace by the presence of oxygen. Under the conditions stated (2/3 oxygen at 490°C for 90 sec), no noticeable damage is done, although some damage does occur after long use. Studies with other strong oxidants (HNO_3, $HClO_3$, and H_2O_2) showed much more severe damage and moreover contaminated the sample with copper traces.

Precision data were calculated for samples with known contents of 15 and 25 ppb and gave RSD values of 12% and 8%, respectively. Limits of detection were the same for the oxygen mixture as for nitrogen alone.

This work raises the possibility of determining volatile elements such as Pb in organic matrices by this technique.

Serum Lithium Determinations Using Flameless Atomic Absorption Spectroscopy

D. T. Stafford and F. Saharovici, *Spectrochim. Acta.* **29B,** 277 (1974)

Flameless methods, particularly those using the carbon rod atomizer, are attractive for the analysis of serum because only small samples are needed and because the methods are highly sensitive. The purpose of this paper is to compare flame and carbon rod atomization.

The analyte was lithium, in serum. Various volumes of the serum sample, 1 μl and larger, were run and compared with results by flame. The results were all in agreement, so there was no point in using carbon rod volumes greater than 1 μl.

Relative standard deviations of 4 to 6% were obtained, and the response was linear, at 670.7 nm, up to a lithium concentration of 2.0 mEq/liter. No interferences were observed.

Comments

Maessen and Posma (250) have also reported on lithium determination with the furnace. Additional papers on the analysis of serum by the furnace technique report on copper (251), chromium (252), iron (253), and aluminum (254).

The Rapid Determination of Barium in Bone by Atomic Absorption Spectroscopy

H. Kawamura and G. Tanaka, *Spectrochim. Acta,* **28B,** 309 (1973)

Bone, which is composed primarily of calcium phosphate, presents some severe problems when the barium content must be determined. The problems arise because of the low concentrations of barium in bone (a Ba: Ca ratio of about 10^{-5}), and because of spectral interference of calcium at the 553.6 nm barium line. Furthermore, chemical separation of small amounts of barium from a calcium matrix is difficult.

A method found practicable involves the coprecipitation of barium and strontium in the presence of calcium by fuming nitric acid. Tests were made to establish the optimum acid concentration and completeness of precipitation by a radiotracer technique.

It was found that optimum concentration was 17.8 N, and that two to three precipitations are needed to reduce the calcium content to manageable levels. Retention of the barium in the combined precipitate with strontium was 98.5%. A few milligrams of calcium remain in the precipitate, but this is tolerable. Strontium does not interfere in the flame determination. The gas used was nitrous oxide–acetylene.

Determination of Heavy Metals in Meats by Atomic Absorption Spectroscopy

S. Slavin, G. E. Peterson, and P. C. Lindahl, *At. Absorpt. Newsl.*, **14**, 57 (1975)

Metals of interest in determinations by the flame method were chromium, iron, and zinc. Cadmium, cobalt, lead, and nickel were determined by the furnace method because concentrations were too low by the former method. Copper and manganese were determined by both methods because certified values were available in the NBS Standard SRM1577 (bovine liver) and so could be used as a check on accuracy. A deuterium background corrector was used with the graphite furnace.

Severe contamination problems were encountered, so that it was necessary to wash glassware in nitric acid and also acid-wash all filter paper. Samples treated were beef liver, beef muscle, and turkey muscle. For cadmium, an electrodeless discharge lamp was used; for other flame determinations the lamps were hollow cathode.

By the flame technique, most determinations fell within about 2% precision. By the furnace technique, precision was about 6%, although several results were much poorer.

Comments

Reference 255 treats the same subject.

Determination of Cadmium in Blood and Urine with the Graphite Furnace

F. C. Wright and J. C. Riner, *At. Absorpt. Newsl.*, **14**, 103 (1975)

A simple, direct method is described. After dilution, the samples were pipetted (50 μl aliquots) into a graphite furnace and atomized.

The standard solution was cadmium succinate in water. The reason for using this compound was that it duplicated the atomization temperature of blood and urine. This temperature was 2150°C, with the graphite tubes cleaned at 2600°C.

Cadmium detection limit by this procedure was found to be 0.5 ppb. Standard deviation of the succinate standards was 1.5%, but actual unknowns were about twice that.

Comments

The analogous determination of lead in blood and tissue is the subject of references 256 to 261.

Trace Elements in Bovine Liver Using Solid Samples in a Furnace

C. J. Pickford and G. Rossi, *At. Absorpt. Newsl.*, **14,** 78 (1975)

In biological analyses below the ppm level, wet-ashing techniques are a constant source of contamination because of the large excess of acid added to the sample.

To avoid the danger of contamination, it was decided to atomize samples directly in the graphite furnace, and furthermore, in order to avoid comparing with aqueous standards, which give poor precision, the method of standard additions was tried.

The sample studied was NBS bovine liver standard SRM-1577, for the certified metals manganese, copper, and lead and for uncertified silver. Samples (4 to 5 mg) were introduced into the graphite tube by means of a tantalum spoon after both windows had been removed to prevent the flowing gas from blowing the light powder out.

The samples were charred at 270°C, and then 5 μl of the addition standard was added by means of a pipet. Normal drying, charring, and atomizing cycles then followed. After the recording of peak heights, the carbon residues were removed from the graphite tube, which was further cleaned by heating to maximum temperature.

Results plotted conventionally for the standard additions method gave straight lines. Concentrations for the four metals ranged from 210 down to 0.06 μg/g, and the RSD averaged about 10%.

An Indirect Determination of Serum Chloride by
Atomic Absorption Methods

H. Bartels, *At. Absorpt. Newsl.*, **6**, 132 (1967)

The chloride content in serum samples was precipitated by silver nitrate, then centrifuged, and the supernatant liquid containing the excess silver made up to standard volume and aspirated.

For 157 replicates, the RSD calculated was 0.52 to 1.18%. Checked against a standard mercurometric method, to guard against systematic errors, the greatest difference for many trials was 1.08% of the amount present.

Centrifugation is much more rapid than dissolving the AgCl precipitate in a solvent and aspirating. It is also much faster than filtering. Results, without the concentrating step of treating the precipitate, were perfectly adequate for biological testing procedures.

Application of Graphite Furnace Atomic Absorption to the Determination of Impurities in Uranium Oxide without Preliminary Separation

G. Bagliano, F. Benischek, and I. Huber, *At. Absorpt. Newsl.*, **14**, 45 (1975)

The authors' laboratory (The International Atomic Energy Agency (IAEA)) is required to determine 10 to 20 elements in their samples of fissile grade uranium oxide in the range 0.01 to 1%. Although the flame technique for the purpose has been described in the literature, the furnace has certain apparent advantages. Small samples can be used, lessening the danger when spills occur, and the furnace can be fitted in a glove box because of the low heat emitted. Thus, a test of the feasibility of the furnace, without preliminary preparation, appeared worthwhile.

Evaluation of the procedure was done on V, Cr, Mn, Fe, Co and Mo. Standard solutions were prepared from uranium oxide of known composition from the New Brunswick laboratory of the AEC and then doped with these metals.

Poor results were obtained from new graphite tubes, but the tubes gave consistent results after the tenth firing. No explanation for this was found. Problems arising in the determination of the six elements are described. The test results for Cr, Mn, Fe, and Co were well within the precision required. For V and Mo results were unsatisfactory, although of some use as semiquantitative; this is ascribed to their refractoriness, so that memory effects cannot be eliminated.

AAS can be substituted in part for the emission spectrograph, but it cannot do qualitative analyses on samples of unknown history.

Analysis of Iron Ores by Atomic Absorption Spectrometry after Pressure Decomposition with Hydrofluoric Acid in a PTFE Autoclave

M. Tomljanovic and Z. Grobenski, *At. Absorpt. Newsl.*, **14**, 52 (1975)

The intention of the authors of this paper was to show the advantages of decomposing ore samples with HF under pressure in an autoclave, compared to conventional procedures. The ores were limonite, hematite, and siderite, in which 18 common elements were determined, most by flame and some by graphite furnace. Comparison standards were reference material from the German Iron Smelting Society.

The finely ground sample, weighing 0.20 g, was transferred to a PTFE autoclave [developed by Langmyhr and Paus (262)] and treated with a mixture of HCl, HNO_3, and HF at 145 to 155°C and about 8 atm pressure for about 30 min. The clear solution was then made up to standard volume and subjected to flame or furnace analysis.

Agreement with the reference samples was good, good enough for routine smelter analysis. In the autoclave, silicon cannot be lost, and samples are brought into solution very quickly. Also included in the analyses was the Fe content, in the range 32 to 59% Fe. The autoclave procedure has become a routine operation in the metallurgical laboratory.

Ion Exchange Separations and Atomic Absorption Determinations of Trace Metals in Ores after Basic Fusion

T. W. Freudiger and C. T. Kenner, *Appl. Spectry.*, **26**, 302 (1971)

Fusion of a mineral sample, customarily in Na_2CO_3, requires 8 to 10 times the weight of the fusion salts as the sample. This high concentration of extraneous salts causes light scatter in the flame and possible clogging of the aspirator. Sodium can be eliminated by separation in an ion column using amino-diacetate chelating resin (Chellex 100), which retains di- and trivalent metal ions above a pH of 6.0 and does not retain the alkali metals. The retained analytes can then be removed from the resin by a wash of dilute HCl.

The powdered sample was mixed with Na_2CO_3 and KNO_3 and then fused in a platinum crucible. The fusion was dissolved in HCl, the pH adjusted to 6 to 8 with NaOH and passed through the ion column. The resin was ashed and the analytes removed by a final passage of 50 ml and $3N$ HCl. This was then diluted to 100 ml and constituted the working solution for conventional flame AAS.

The elements of interest were Co, Cu, Mn, Zn, Fe, and Pb, and the question to be determined was the degree of retention by the resin. For this purpose four NBS samples were run through the procedure—burnt magnesite, plastic clay, antimony ore, and zinc ore. All of the metals mentioned above were retained at least to the extent of 95%. Chromium proved to be unsatisfactory, in part because it was present as the chromate.

Compared to direct analysis of the fusion solution, precision was better by a factor of five when the eluted solution was used.

Comments

This is a classical application of ion column separation. The sample solution, stripped of its unwanted constituents, gave a marked improvement in results. However, the ion column method is slow and not very suitable for a large number of samples.

Korkisch and Gross (263) also used ion exchange on minerals.

Determination of Trace Elements in High-Purity Aluminum by Atomic Absorption Spectroscopy (in German)

R. Höhn and E. Jackwerth, *Spectrochim. Acta,* **29B**, 225 (1974)

The objective in this work was to use a large sample (10 to 25 g) in order to obtain sufficient amounts of the analytes to produce a good signal, and at the same time to avoid both the high concentration of aluminum salts in the aspirated solution and dilution of the sample. Neither condition would be conducive to good trace metal determination.

The separation procedure finally worked out was simple and rapid. It consisted in dissolving the aluminum in HCl while in contact with metallic mercury, which precipitates the traces by electrolytic action and simultaneously dissolves them to form amalgams. Bi, Cd, Ga, In, Tl, and Zn can be recovered by this process, with recoveries better than 95%.

The procedure was tested for the levels of recovery in the following manner. Samples of analyzed standards were first coated with about 1 g of mercury, then dissolved in HCl at two strengths, $1.5N$ and $6N$, in a volume of acid of 100 ml. The dissolution was stopped when all but about 10 to 30 mg of the aluminum had gone into solution. The dissolved aluminum was discarded by careful decantation, followed by two to three washes. The residue was dissolved in concentrated nitric acid, the excess acid removed by evaporation and the metal salts dissolved in about 50 ml of water. The mercury was recovered from this solution by precipitation with about 5 ml of formic acid. The solution, now free of mercury, was diluted to standard volume and analyzed by conventional AAS procedure.

Recovery in the $6N$ HCl was at least 95% complete for all metals listed, with the exception of zinc, which was poor. However, in the weaker acid all the listed metals, including zinc, were recovered to better than 95%.

The relative standard deviation of the procedure averaged about 5%, and detection limits were in the range 0.02 to 2.0 ppm.

Comments

This is a clever application of electrolytic precipitation, an old and familiar reaction. However, it applies only to metals that readily form amalgams.

Burke (264) also discusses the determination of trace elements in metals.

Determination of Bismuth in Nickel-Base Superalloys by Atomic Absorption Spectrophotometry

J. A. White, Sr., W. L. Harper, Jr., A. P. Friedman, and V. E. Banas,
Appl. Spectry., **28**, 192 (1974)

The presence of bismuth in nickel- and cobalt-based high-temperature alloys, down to a concentration of 1 μg/g, causes certain deleterious effects on the physical properties of the alloys. An analytical method capable of measuring bismuth to a concentration of about 0.5 μg/g was therefore desired.

The procedure for the solution of this problem is based on an electrolytic separation of the bismuth from the matrix metals. Concentration of the analyte into a much smaller volume was also effected. The 8 g sample was dissolved in a mixture of HF, HNO_3 and HCl in a Teflon beaker, and after dissolution and dilution with water, hydrazine dichloride was added to prevent bismuth complexes.

The platinum gauze cathode was prepared by plating with a layer of copper and then with the analyte solution to which some copper solution had been added. The cathode was then stripped in a measured volume of a mixture of HNO_3 and H_2O_2 and aspirated into an acetylene–air flame. Comparison was against synthetic standards containing the same concentrations of acid and peroxide as the unknown electrolyte.

In this process the question of loss of analyte metal was important. Tests were made by spiking a solution of the matrix metals and running through the procedure. It was determined that recovery was better than 99%. A reproducibility study showed that the standard deviation was 0.1 μg of bismuth in the range 0.1 to 1.0 μg/g.

Comments

This paper illustrates the effectiveness of electrolysis as a separation technique and the possibility of concentrating the wanted element into a small volume. Furthermore, electrolysis is an old and well-tested method, so little if any preliminary testing is needed.

Atomic Absorption Spectrometry of the Lanthanides in Minerals and Ores

W. Ooghe and F. Verbeek, *Anal. Chim. Acta,* **73**, 87 (1974)

Standards were prepared by solution of the lanthanide oxides in 6M hydrochloric acid, evaporated to dryness over a heat lamp, and taken up in weak HCl. A buffer of KCl was added to suppress ionization.

Tests to establish optimum parameters resulted in choosing a KCl concentration of 20 g/liter in 80% methanol, with a maximum concentration of the sample at 30 g/liter.

Powdered samples of bastnasite and gadolinite (0.5 to 1.5 g) were dissolved in $6M$ HCl; monazite was dissolved in perchloric acid to which a few drops of HNO_3 had been added. Excess acid was removed by evaporation and baking. The residue was then redissolved in minimum acid, and precipitated SiO_2 was removed by filtration. The solutions were made up to 50 ml, and this constituted the working solution.

Aspiration and atomization were conventional—in a laminar flow burner and a nitrous oxide–acetylene flame. The presence of the methanol increased the sensitivity markedly. The working curve was linear up to an absorption of 25%, and sensitivity, in μg/ml for 1% absorption, varied from 14.8 for lanthanum to 0.09 for ytterbium.

The authors present a list of preferred lines for the analyses, but they have apparently been taken from the Perkin-Elmer "Cookbook": the list is identical to the list in Table 5.1 of this volume, which has also been taken from that source. The only spectral interference mentioned is that of the line Nd 492.4, with a line of praseodymium, a common associate of neodymium in minerals.

As the possibility of matrix mismatch existed, the calibration curve was in some instances established by the method of standard additions, although this was unnecessary for samples of similar origin; for these a comparison standard was used.

As a check against systematic errors, the results by AAS were compared to two other methods and found to be satisfactory, although no arbitrary figures were stated.

Comments

The method presented here is conventional but shows that elaboration in preparing this group is not necessary. Twelve rare earth elements were treated; only the very scarce ones were omitted. The not-so-rare cerium cannot be determined by direct AAS, as there are no suitable lines. However, an indirect method has been suggested (225), based on the formation of molybdocerophosphate and evaluation of the molybdenum.

Additional references pertaining to the determination of the rare earths are given in 265 to 269. A review of the analysis of the rare earths has been published by Thomerson and Price (270).

Atomic Absorption Spectrophotometric Analysis (of Silicate Minerals) by Direct Atomization from the Solid Phase*

F. J. Langmyhr and Y. Thomassen, Z. *Anal. Chem.*, **264,** 122 (1973)

Silicates were atomized directly in a graphite furnace heated by high-frequency induction. The maximum temperature attained was about 2000°C.

The samples were gneiss, granulite, schist, anorthite, peridotite, and granite. The analytes were cesium and rubidium. The sample weights varied from 1 to 20 mg. For comparison, the samples were decomposed by treatment with HF and H_2SO_4, taken up as solutions, and run by conventional flame technique. The cesium content by flame determination was too low to obtain a signal, but rubidium comparisons were possible.

No interferences from Al, Fe, Ca, Mg, Na, or K were found when the signals were integrated, but peak signals were reduced by the presence of Na and K. Anions, however, do interfere. SO_4 gave the best signal when integrated. The authors suspect that Rb enters the gas phase as the molecule, hence the lower signal.

The precision was found to be good by furnace (5 to 15%), slightly less than by flame but equal to the usual furnace results. For cesium at about the 1 ppm level, the errors were 10 to 20%.

Comments

The authors are fully aware of possible sampling errors because of the small sample size. They assumed, however, that cesium and rubidium replaced the potassium in the minerals, so the dangers of segregation were greatly reduced, as potassium in these minerals was at a much higher concentration than cesium and rubidium. They referred to the paper by Wilson (109), who had made a study of segregational errors with respect to sample size versus fineness of the sample.

Note, however, that for samples in which the analytes are not present in solid solution, but in separate minerals—that is, in discrete particles that may or may not appear in the small sample—segregational errors in either direction may be so high as to make the technique impractical.

*Original paper in English.

Furthermore, it should be pointed out that direct atomization of the solid sample loses all the advantages of the solution technique: ability to change analyte level by dilution, possibility of removing interfering substances, and the possibility of using a large sample to concentrate the analytes. Also, segregation in solutions is entirely eliminated.

Langmyhr and co-workers (271) have recently published a series of papers on direct atomization of various sample types; the interested reader is referred to these papers. See also p. 132 in this book.

A Comprehensive Scheme for the Analysis of a Wide Range of Steels by Atomic Absorption Spectrophotometry

D. R. Thomerson and W. J. Price, *Analyst,* **96,** 825 (1971)

This long paper describes the determination of twelve constituents commonly found in steel: Mn, Ni, Cr, Mo, Cu, V, Co, Ti, Sn, Al, Pb, and W. All, except for the last, are based on a single 1 g sample dissolved in perchloric acid. Tungsten, which is not retained in solution, is determined after a separate dissolution of the sample.

Perchloric acid as the final solvent was chosen because it causes the least interference of all the common acids. The only addition necessary is iron from a stock 5% solution derived preferably from the pure metal. The 1 g unknown sample is dissolved in a mixture of hydrochloric, nitric, and perchloric acids and evaporated to fumes. After cooling, the salts are dissolved in water, filtered, and made up to standard volume. The same procedure is followed for the standard solutions, using pure iron.

If the tungsten content is higher than 0.5%, a slightly different procedure is followed for tungsten and molybdenum. Higher concentrations for all the metals may be diluted, but the final aspirated solution should contain 1% iron.

This scheme was tested for all 12 metals listed on a series of steels, from mild to high-alloy and stainless types, of the British Chemical Standard Steels. Results were in excellent agreement with certificate values, and in most cases the replicate results had a smaller spread than the spread reported for the BCS samples.

Silicon, while not included in the scheme, can be determined by the traditional gravimetric method on the precipitate made insoluble after acid fuming. Thus, no separate sample is needed for the silicon.

The paper contains a discussion for each metal treated, with literature reference of previous work. Also included is a table showing the composition of standards with respect to the matrix metals and the metals to be determined with these standards.

Determination of Beryllium in Environmental Materials by Flameless Atomic Absorption Spectroscopy

J. W. Owens and E. S. Gladney, *At. Absorpt, Newsl.*, **14**, 76 (1975)

The need for monitoring beryllium in the environment has led to a search for methods that are sensitive and easily applied. Most methods for the determination of beryllium in the past have been insufficiently sensitive or have required lengthy preparation. The method described here uses the graphite furnace for high sensitivity and requires little preparation for both silicate and biological samples.

The working solution is prepared either by an acid dissolution for silicates or by wet-ashing for biological material. The final form is as the perchlorate.

The test samples were NBS reference materials Fly Ash, Coal, Orchard Leaves, and Bovine Liver, with beryllium content ranging from 12 to 0.005 ppm. Agreement with the certified values was very good, proving that the acid treatment was practicable and that sensitivity was more than ample.

Attempts to run the samples directly in the graphite furnace, either in a tantalum boat or dumped in the tube, were not successful, owing to several difficulties. The tantalum boat became brittle at 2700°C, which was the temperature needed to clear the furnace of memory effects, which required that the furnace be baked after each run. Fine powders tended to blow out of the tube.

Direct Determination of Calcium in Natural Waters

D. A. Ward and D. G. Biechler, *At. Absorpt. Newsl.*, **14**, 29 (1975)

Samples were diluted 1:1 by simultaneous aspiration of the sample and a solution containing 2000 μg of sodium, with mixing in the nebulizer.

Atomization was in a nitrous oxide–acetylene flame in a standard 2 in. premix burner.

The Ca line at 422.6 nm was found to be free of interference from K, Mg, Al, Fe, Si, PO_4, and SO_4. Agreement with EDTA titrations and the method of standard additions was excellent. Forty samples per hour could be put through. It was not necessary to measure the volume of the sodium suppressant solution aspirated, only to ensure reproducibility.

Comments

A paper on trace metal analysis in seawater has been published by Segar and Gonzalez (272), and Ediger has reviewed the analysis of water by AAS (273).

Arsenic Determination in Tobacco by Atomic Absorption Spectroscopy

H. R. Griffin, M. B. Hocking, and D. G. Lowery, *Anal. Chem.*, **47**, 224 (1975)

Determination of arsenic is of interest from an environmental viewpoint because arsenical pesticides are sometimes used on growing plants. Neutron activation gives reliable indications but has been used for few samples on account of the elaborate equipment and operator skill needed. An equally reliable AAS method should result in much wider application.

In the present method the tobacco samples are digested in a Kjeldahl flask with HNO_3 and $HClO_4$, with heating to 200°C. After cooling, the contents are diluted to a standard volume and an aliquot is removed. In this aliquot, arsine gas is generated with zinc in the usual manner. The arsine generated is frozen out by cooling in liquid nitrogen; the tube, still at low temperature, is connected to the aspirator of an AA instrument, the closed tube allowed to reach room temperature, and the stopcock opened to allow the arsine to pass into a flame and be measured.

The working curve was made on known concentrations by this procedure. The wavelength used for the arsenic determination was 193.7 nm, and the recommended flame is argon–hydrogen with entrained air, although nitrous oxide–acetylene with entrained air is also practicable. The detection limit was about 0.6 μg of As and the RSD varied from 4% to 18%. The working curve was linear to 12 μg As. A blank on the zinc must be run.

An advantage of the freezing-out technique is that it permits the treatment of large samples when the arsenic content is very low.

Also reported are many analyses of arsenic content in tobaccos grown in many parts of the world. This presence of arsenic should be of concern to environmentalists because of its possible carcinogenic properties.

Comments

The concentration of other volatile elements may be done by this freezing-out technique.

A paper describing a digester for leaves has been published by Griffin and Hocking (274).

Determination of Selenium in Water and Industrial Effluents by Flameless Atomic Absorption

E. L. Henn, *Anal. Chem.*, **47**, 428 (1975)

Selenium is a toxic, cumulative substance. The traditional method used for its determination is colorimetric, which is slow, difficult, and not very reliable. The conventional AAS procedure suffers from low sensitivity.

The present method uses a furnace for volatilization and an electrodeless lamp for excitation. Experimental conditions are as follows:

Wavelength: 196.0 nm with 2 nm bandpass
Sample volume: 100 μl
Furnace temperature: 2500°C

Tests for interference showed that a list of 31 metals caused moderate to severe interference. As the sample was given a preliminary digestion with H_2O_2, this treatment converted the Se to the anion so the interfering cations could be removed by ion exchange.

The high acid content of the sample after removal of the cations still caused severe depression of sensitivity. A search of reagents whose addition would improve sensitivity led to molybdenum, which was found to have an optimum effect at a concentration of 30 μg/ml. The presence of Mo also permitted a much higher charring temperature before Se was lost.

The ion-exchanged solution, with Mo added, yielded a sensitivity of about 1 μg/ml, with a linear calibration curve to 50 μg/ml and a coefficient of variation of 7.5%.

Comments

This is a long paper, and the various steps in the procedure were thoroughly tested.

Pierce et al. (275) have also published a paper on the analysis of waters for selenium and arsenic.

Analysis for Chromium Traces in Natural Waters

J. F. Pankow and G. E. Janauer, *Anal. Chim. Acta*, **69**, 97 (1974)

The toxicity of chromium as the chromate is well known, but its presence in natural waters is largely unknown. As the concentration in such samples is extremely low, an efficient concentration technique is called for. An ion-exchange column proved effective.

The resins used were research grade XN-1002 and production grade IR-900, made by Rohm and Haas, and resin No. AG1-X4, a strongly basic exchanger from the Bio-Rad Laboratories. The best way of using these resins was studied in connection with standards at concentrations equivalent to the expected levels in natural waters. The standards were adjusted to pH 5.0 with HNO_3 and 20 ppm SO_4 added to simulate the average composition of natural waters.

The wavelength chosen was at 357.7 nm, and a reducing (red) flame was used (probably nitrous oxide–acetylene, but this was not mentioned). This procedure made possible the determination of 0.1 ppb of chromium as the chromate starting with a 1 liter sample.

The bulk of the paper describes the various tests of the ion-exchange procedure and reports on the concentration factors and degrees of retention. The exchange can be followed visually because a yellow ring forms in the column near the end of introduction of the sample solution. To keep the exchanged volume small, the preferred flow of sample through the column was upward.

The Determination of Mercury in Air Samples by Flameless Atomic Absorption

J. F. Lech, D. D. Siemer, and R. Woodriff, *Spectrochim. Acta,* **28B**, 435 (1973)

The usual methods of determining mercury in air for environmental control have the disadvantages of either being comparatively insensitive or requiring some sample preparation, with danger of loss. The direct flameless technique described in this paper is extremely sensitive and simple.

The method consists in filtering a small volume of air through a porous graphite cup whose inner surface has been coated with a thin layer of gold formed by electrolytic deposition. An air sample drawn through this surface catches both particulate mercury compounds and the vapor. The metal can then be determined by conventional flameless techniques.

The cups, the special holding device, and their operation have been described in an earlier paper by Woodriff and Lech (276). A layer of gold weighing 10^{-3} to 10^{-4} g is deposited from a chloride solution by a direct current (with the cup as cathode) obtained from a battery. The aspirating holder is formed from methacrylate plastic, which is easily cleaned. Suction is provided by a 50 ml syringe, which is easily operated in the field.

The cup with sample is then placed in a tube furnace and heated to 900°C, which drives off all mercury in both the molecular phase and the amalgam. The mercury is measured conventionally with a mercury hollow-cathode lamp, using the 253.65 nm line.

Standards were prepared by two methods: (1) by adding known quantities of gold in solution directly into the cup and drying under a heat lamp and (2) by filling a flask with mercury vapor (from the metal), equilibrating the vapor at a fixed temperature, and then drawing a measured volume through the cup. Concentration of the vapor can be calculated by applying the gas law.

Tests to ensure that retention was complete were made by passing the stripped gas into a long absorption tube and measuring the degree of absorption of a beam from a mercury lamp. Unplated cups showed a retention of only 10%, while plated cups retained 99% of the analyte.

Precision measurements gave a relative standard deviation of 11.2%, with a detection limit of about 10^{-10} g of mercury. On the real samples the

RSD was about 30%. Sensitivity of the method was far above safety limits, even for small air sample volumes.

Comments

Woodriff and co-workers have published several papers describing the porous cup application to other metals of interest to investigators of air contamination. The literature contains many papers on the determination of mercury in air; some are listed under references 277 and 278, and reviews under references 279 to 283.

Determination of Platinum and Palladium in Biological Samples

R. G. Miller and J. U. Doerger, *At. Absorpt. Newsl.*, **14**, 66 (1975)

With the present use of platinum and palladium as catalysts in automobile exhaust systems, a possible environmental problem has arisen. Whether the two metals will act as pollutants or not, it seems that methods are needed to determine the lowest possible detection limits of these metals.

As no overall procedure for routine analysis of Pt and Pd in biological materials exists, parts of other methods have been utilized. Samples were slowly wet-digested in HCl and aqua regia, to make sure that the metals would dissolve, whatever form they were in. Nitric acid was removed by evaporation with HCl.

The absorption measurements were made with a graphite furnace plus a background corrector. The atomization temperature was 2700°C for Pt and 2600°C for Pd.

The resulting calibration curves on 20 repeat runs of each of four synthetic samples was a straight line passing through the origin. The realistic lower limit of detection of Pt and Pd in 1 g of tissue is 0.20 μg. The sensitivity at 1% absorption is 1175 pg for Pt and 160 pg for Pd. Recovery of both metals was about 95%.

Comment

See also Macquet and Theophanides (284).

Atomic Absorption Spectrophotometric Determination of Lead in Beverages and Fruit Juices, and of Lead Extracted by their Action on Glazed Ceramic Surfaces

D. Gehiou and M. Botsivali, *Analyst,* **100,** 234 (1975)

Samples of beverages for determination of lead content are prepared by evaporation of the liquid, followed by ashing of the residue, according to official AOAC methods. Gehiou and Botsivali sought to test the feasibility of aspirating the samples directly into a flame and measuring the absorption.

The test beverages were orange juice, Coca Cola, instant coffee, tea, and dry wine. Included for comparison was a 4% acetic acid solution. Standards were prepared of lead nitrate solutions in water, in a water–ethanol mixture, and in 4% acetic acid. It was found that aspiration rates differed for the various sample types, so that a correction had to be made.

By direct aspiration, lead could be determined easily down to about 0.1 ppm, although sensitivity was not as good as by the ashing technique. Interference in most sample matrices was negligible. The standard deviation found for the range 0.5 to 2.0 ppm was 0.04 ppm.

Leaching rates of Pb from glazed surfaces, such as are found on cups, were measured for common beverages and the acetic acid solution, and the results are presented in a table.

Limits set by regulatory authorities for the presence of lead in beverages are, by British Standard Limits, 0.2 ppm in nonalcoholic beverages, 0.5 ppm in fruit juices, and 1.0 ppm in wines. The U.S. Food and Drug Administration limits leachable lead in ceramic ware to 7 ppm. Sensitivity by direct aspiration is well within these limits; besides, the method is so simple and rapid as to be advantageous over ashing.

Comments

See also the paper by Krinitz and co-workers (285).

While the present authors were concerned with the leaching of lead from glazed vessels by certain beverages, their suggestion that the beverages could be atomized directly could be applied to the general analysis of beverages.

A Method for the Determination of Cadmium in Plant Material by Atomic Absorption Spectroscopy

Duane Boline and W. G. Schrenk, *Appl. Spectry.*, **30**, 607 (1976)

The official AOAC method for determining cadmium in plant material is by colorimetric comparison. In an earlier paper, Kahn et al. (286), who used an aqueous solution and flame for atomic absorption, reported that they experienced some interference. This paper reports work to minimize interference and to increase sensitivity.

Preparation consisted in ashing a 2 g sample of ground-up plant material by both dry- and wet-ashing techniques. For the former, the sample was heated overnight in a dish at 500°C. For the latter, the sample was treated with HNO_3 diluted 1:1 plus $HClO_4$ (70%) and boiled to a yellow straw color. The solution was then evaporated to dryness. Both this residue and the ignited ash were dissolved in 10 ml of $1N$ HCl and diluted to 25 ml with methanol. This was aspirated into a flame of air–acetylene. Standards were also diluted to 60% methanol. A blank must be run on the methanol. The range of standards was from 0.05 to 2.0 μg/ml of cadmium.

Removal of the interference found by Kahn et al. was ascribed to the presence of alcohol in the sample solution. Of four alcohols tested, methanol gave the best sensitivity. Interferences by Ca, Na, K, Cu, Fe, PO_4, and SO_4 were investigated and found to be without effect.

Completeness of cadmium recovery was tested by an addition method and found to be 87 to 104%, with no apparent difference between wet and dry ashing. Flash vaporization (by resistance-heated tantalum boat) of the solution revealed several objections: the solution tended to creep over the tantalum surface, several peaks were observed, and reproducibility was too poor to be practicable.

The authors concluded that atomic absorption by air–acetylene flame gave results that were as good as or better than the official AOAC method, besides being simpler and more rapid.

Comments

Note that dry ashing resulted in no loss of cadmium. The conventional wisdom holds that an element so low in melting point cannot be dry-ashed without loss.

Indirect Determination of SO₄ in Water by Atomic Absorption

O. K. Galle and L. R. Hathaway, *Appl. Spectry.*, **29**, 518 (1975)

A barium chloride solution of known concentration is added in excess to the unknown solution, and the remaining Ba is measured by AAS.

As applied to natural waters, the acidity is adjusted to $0.02N$ with HCl, the $BaCl_2$ solution is prepared with five times its concentration with NaCl (to suppress ionization in the flame), and measurement is at a wavelength of 553.5 nm, with background correction.

The precipitation vessel is a volumetric flask, which is thoroughly shaken after addition of the sample and precipitating solution and allowed to stand for 12 hr or overnight. The supernatant liquid is then aspirated without filtration into a nitrous oxide–acetylene flame. Comparison is with a set of standards prepared with Na_2SO_4 and carried through the same procedure.

Agreement with gravimetrically determined samples was quite good, and a precision test on 20 samples ranging in content from 13 to 500 ppm of SO_4 gave an RSD of 20% for the lowest and 1% for the highest concentration. Linearity of the working curve ranged from zero to 120 ppm.

Comments

This procedure is typical of the whole class of indirect methods in which the excess of precipitant is measured. Obviously, it can be applied to other common anions—Cl, Br, I, PO_4, and others. Mixtures in which the precipitating cation reacts with more than one analyte would cause difficulties and may require a prior separation. Effluents and brines may need filtration.

Optimum conditions for the precipitation of other cations (Ag for the halides, and Mg for PO_4) have long since been worked out for the gravimetric method; a standard text should be consulted.

Additional references for SO_4 are listed under 287 to 289 and a review of the subject under 290.

Determination of Water-Soluble Sulphate in Acidic Sulphate Soils by Atomic Absorption Spectroscopy

J. A. Varley and Poon Yew Chin, *Analyst*, **95**, 592 (1970)

Soil extracts with widely varying SO_4 contents can be accommodated to AAS analysis by making a preliminary conductivity measurement to establish approximate content and then taking an appropriate aliquot. The concentration of the barium chloride solution was adjusted to give a full-scale deflection when aspirated directly; hence any reduction in reading would still fall on the scale.

The efficiency of the precipitation step was tested and found to be about 98% in recovery of the barium under the conditions used.

Possible interferences were investigated, and none were found for Na, K, Mg, Mn and Fe. Calcium and phosphate do interfere; higher readings for SO_4 begin when the Ca content equals the sulfate content, and for PO_4 significant errors begin at about one-fifth of the SO_4 content.

The method was checked against traditional gravimetric and turbidometric procedures, and results checked well.

Indirect Determination of Tantalum by Atomic Absorption

Tsutomi Matsuo, J. Shida, and S. Kudo, *Bull. Chem. Soc. Japan*, **46**, 3595 (1973)

A report on the determination of tantalum, as a complex with molybdenum and using molecular absorption techniques, led the authors of the present paper to apply the molybdenum-complexing reaction to an indirect AAS method, determining the tantalum indirectly by determining the molybdenum.

The procedure that was developed was to add ammonium molybdate to the tantalum solution (either the sulfate or the chloride), extract the molybdotantalate into MIBK, wash to remove excess of reagent, and then aspirate the MIBK solution directly into the flame.

Acidity for the organic extraction should be in the range 0.5 to $0.7M$ HCl. Concentration of the molybdate solution should be about $0.2M$, to which the tantalum standards are added. Formation of the complex is slow; about 15 min must be allowed for the reaction before extraction. The MIBK extract is relatively stable.

The calibration curve is linear o $2.0 \times 10^{-5}M$ tantalum. Interfering elements, because they also form molybdate complexes, are arsenic, germanium, phosphorus, silicon, and niobium. Limit of detection was not mentioned, nor was the flame type, presumably because they are based on the molybdenum determination.

An Indirect Amplification Procedure for the Determination of Niobium by Atomic Absorption Spectroscopy

G. F. Kirkbright, A. M. Smith, and T. S. West, *Analyst,* **93,** 292 (1968)

Niobium in solution is complexed by the addition of molybdophosphoric acid, the excess reagent is removed by extraction into isobutyl acetate, and the niobium phosphomolybdate complex remaining in the aqueous phase is then extracted into butanol. The molybdenum contained in the butanol solution is determined by conventional procedure by AAS, at 313.2 nm in a nitrous oxide–acetylene flame.

High sensitivity is achieved by the 11-fold amplification, owing to the Nb/Mo ratio in the compound, by the efficient nebulization of the butanol, and by the opportunity for concentration of the analyte through organic extraction. Details of the separation and extraction must be followed carefully.

The overall RSD obtained by this procedure was 2.4%. The calibration curve with respect to niobium was linear over the range 5 to 50 μg, or, for the concentrations used, 0.22 to 2.2 ppm. Subtracting the blank caused the curve to pass through the origin.

A series of 28 of the common elements were tested for possible interference, and none was found. However, arsenic and germanium, which also form heteropolyacids under similar conditions, can be removed by volatilization of their chlorides, and silicon by volatilization with HF. Titanium, when present in large amount, causes interference, but if the amount is comparable to the niobium it can be tolerated. Vanadium(V) and chromium(VI) can be tolerated up to threefold and 45-fold, respectively. Tantalum does not interfere.

The extraction processes were studied for completion by Babko and Shkaravskii (291).

The Quantitative Estimation of Lead, Antimony, and Barium in Gunshot Residues by Non-Flame Atomic Absorption Spectroscopy

G. D. Renshaw, C. A. Pounds, and E. F. Pearson, *At. Absorpt. Newsl.*, **12**, 55 (1973)

When a revolver or automatic pistol is fired, the hand or hands holding it receive particles of lead, antimony, and barium from the discharge. The antimony and barium come from the priming of the cartridge. Determination of these metals in the past was done by colorimetric methods which have since been shown to be unreliable. Neutron activation is a reliable method but is expensive and slow; it is not applicable to lead determination.

The procedure described in the present paper can be completed in one hour. It is based on the use of the graphite furnace and requires a sample volume of 20 μl.

The hand of the person suspected of firing the pistol—the alleged perpetrator, in police parlance—is swabbed with a cloth moistened with dilute HCl and the residue is then extracted from the cloth with 3N HCl. For the lead determination, half of the sample is neutralized with NH_4OH to a pH of 2 to 3, complexed with a 1% aqueous solution of APDC and collected in 0.5 ml MIBK. A volume of 20 μl is transferred to the furnace and the lead determined by conventional methods. The remaining portion of the MIBK solution is adjusted to a pH of 8 to 9, the barium is extracted by TTA (thenyl trifluoroacetone) in MIBK, transferred to the furnace, and determined.

The other half of the sample HCl extract is treated with ceric sulfate to oxidize the antimony to the pentavalent form and the antimonic chloride is extracted into MIBK, with aliquots taken for the antimony determination.

This procedure gave sensitivities of 10^{-10} g for lead and barium and 10^{-9} g for antimony. Typical gunshot values are in the 10^{-7} g range, so that the method gives results well above detection limits.

Comments

For forensic purposes, the quantitative accuracy of results is quite unimportant; all that is necessary is to obtain an unequivocal (yes or no) answer as to whether the analysis result is above any possible background. A check of the work is impossible, because the sample is used up.

Graphite Tube Study at Gunshot Residue Levels

J. H. Sherfinski, *At. Absorpt. Newsl.*, **14**, 26 (1975)

A common police test to prove that a handgun had recently been fired is to swab the hand of a suspect with an acid-impregnated cloth and then show the presence of barium, antimony, and lead in the swab. Flameless atomic absorption is the method of preference over neutron activation. However, Renshaw (292) questioned the reliability of the barium determination because of possible carbide formation in the graphite tube of the furnace, with only partial volatilization of the refractory carbide.

Sherfinski examined this theory. He used a Perkin-Elmer furnace, Model HGA-2100, in which the purge gas flows into either end of the tube and out through the sample hole in the center. Sample fume would then tend to be carried into the hottest portion and away from the cooler ends.

The barium range for swab tests is typically 0.1 to 1.0 μg, depending on the type of weapon fired and the number of shots. A standard containing 0.5 μg/ml of barium was prepared, with actual volumes of 50 μl charged into the tube. The signal from this weight of analyte was adequately strong.

Repetitive tests of this standard were run up to failure of the graphite tube. The atomization temperature was 2500°C, and each peak was recorded on a stripchart. From this the RSD was calculated.

The tube failed on the 130th run. The RSD increased slowly from about 2% for the initial tests to about 10% at failure. The conclusion, therefore, was that the barium determination by graphite tube furnace is entirely reliable, provided that care is taken to confine the analyte fume to the hot portions of the tube.

Furthermore, the stripchart recording provides a signal of approaching tube failure: the appearance, at about the 70th run, of a transient negative signal that became stronger with succeeding runs. No explanation for this was offered.

Comments

Significance of the work described in this paper goes beyond the forensic field or the determination of barium. The manner of flow of the purge gas appears to be important for avoidance of memory effects. Graphite tubes, even when heated to near the maximum temperature of the furnace, have a comparatively long life, and the negative transient should prove a useful indication.

Analysis of swabs for gunshot residues is only a small part of the work of forensic chemists. They are interested also in such problems as bullet composition, residues from explosives, ashes from drugs, paint chips, and so on, but these problems are also common in industrial, metallurgical, and geological fields, so there is no need to repeat these methods in the forensic section.

Goleb and Minkiff (293) point out that many collection kits used for the storage of swabs contain traces of barium and may cause contamination. The same authors discuss a tantalum strip atomizer (294) for gunshot residue analyses.

Trace Characteristics of Powders by Atomic Absorption Spectrometry

B. V. L'vov, *Talanta*, **23**, 109 (1976)

This is a review of recent Russian work on direct volatilization of solids, and is presented here for those unfamiliar with the Russian literature.

The scheme outlined in this review is a new method of volatilizing solid samples, either by flameless heating or combined with a flame. Two atomizer designs are described. In the first, a graphite capsule (a rod drilled to near the other end to form a cup) that carries the sample is held between two graphite electrodes for the usual resistive heating. The sample chamber is tightly sealed, but the sample vapor is allowed to pass through the porous wall of the capsule into the flame of a Meker burner. The exciting beam, parallel to the capsule, passes just above its wall and through the flame.

In the second design, two graphite tubes, one inside the other, are arranged concentrically with space between them. The sample is placed in this space, the ends of the inner tube clamped between electrodes, and heated (with no flame). The vapors pass through the wall of the inner cylinder and out through its open ends. The sample chamber is sealed by the electrodes. The exciting beam passes through the center of the assembly.

The path of the vapor flow is controlled by coating the appropriate surfaces with pyrolitic graphite, which makes them impervious to gases. Oxidation is minimized by bathing the hot graphite in an inert gas.

The volatilization process was improved by mixing the sample with powdered graphite, an idea borrowed from a common practice of emission workers using the carbon arc. This mixing with graphite provided an opportunity to control in some degree the quantity of sample.

Presented in the review is a table of 27 common elements that lists the detection limits to a 2 σ standard deviation, both in terms of absolute weight of analyte and in percent concentration. Actual sample weights ranged from 10^{-11} to 10^{-13} g. L'vov estimates that at least 50 elements can be determined by this method, using the capsule-in-flame design; fewer by the concentric tube design. A comparison of results with those obtained by emission spectroscopy showed good agreement, with no bias.

Reliability tests showed a relative standard deviation of 4 to 8%. The time required to run through a series of 10 replicates of 10 elements each, including the grinding, mixing, loading, and change of hollow-cathode lamps, was 5 hr. Samples were not weighed; a scoop was used to take a standard volume of the solid. Capsules and tubes were reusable for 20 to 40 determinations.

Comments

This Russian technique of solid sample volatilization can be compared most closely with the work of Langmyhr and co-workers. The advantages over the latter are the use of larger samples (40 against 20 mg), which lessens the sampling errors, and the removal of fume particles from the optical path, which reduces background and provides an unobstructed beam.

Precision and detection limits appears to be similar to the Langmyhr experience.

One would expect that memory effects would cause trouble, but L'vov makes no mention of this, although he used both capsules and tubes repeatedly.

To hasten the procedure, samples of the powders were measured out by volume, not by weight. This could introduce large errors because of changing bulk density. Weighing of individual samples, although much safer, would greatly increase the time per determination.

See other work (295, 296) which makes use of the porosity of graphite.

Precise Calcium Phosphate Determination

M. B. Tomson, J. P. Barone, and G. H. Nancollas, *At. Absorpt. Newsl.*, **16**, 117 (1977)

This paper describes a method for the determination of both calcium and phosphate in the same biological sample. The calcium determination is by conventional flame procedures, but the phosphate procedure presents a novel idea that should have application to other anions.

Standards and unknowns are prepared by adding lanthanum chloride (for ion suppression), ammonium molybdate, and ammonium vanadate as solutions and then diluting to a standard volume. This constitutes the working solution. The yellow color of the vanadium phosphomolybdate is allowed to develop for 2 hr, after which the calcium is determined. For the phosphate determination, a 1 cm absorption cell, such as is used in molecular spectrophotometry, is filled with the solution and clipped to the cooled burner head, and the absorption is measured by using the same wavelength of the hollow-cathode lamp as is used for the calcium measurement. This wavelength is 422.7 nm, which is very close to the peak absorption at 420 nm of the yellow solution.

The calcium calibration curve showed a distinct curvature, which was ascribed to concentrations above the linear range. The phosphate determination required a careful addition of the color reagent to within about 2% for each sample. The RSD for calcium over the concentration range varied from about 0.20 to 0.32%. The phosphate absorbances were within one digit in the third decimal place.

The authors state that the advantages in using a modern AA spectrophotometer are that the cross-sectional area of the hollow-cathode beam is larger than the beam in the usual molecular spectrophotometer, which reduces the effect of any dirt or imperfection of the cell walls, the cell is readily accessible, the light is highly achromatic and reproducible, and the electronics are very stable.

Comments

What the authors describe for their phosphate determination, without using the term, is a high-quality photoelectric colorimeter. The only change needed from an AA instrument to a colorimeter is to substitute an absorption cell for the burner. Thus, at trifling cost, colorimetric methods may be used for those elements, both cationic and anionic, that cannot be determined, or only very poorly, by conventional AA. Examples of the latter are such metals as niobium, tantalum, cerium, tungsten, and uranium.

1

COMMERCIAL EQUIPMENT

Lists of manufacturers, together with descriptions of their wares and their prices (to about 1970) have been published in a chapter of the book by Dean and Rains (104) and in the handbook by Veillon (40). These two references cover only instruments made in the West; it proved impossible to obtain the corresponding information from the communist bloc nations, although L'vov (79) makes frequent mention to AAS equipment made in the USSR.

Veillon lists 17 manufacturers of complete instruments and accessories whose products either are made in the USA or can be obtained here. Of these, by far the largest share of the business in this country is in the hands of four American firms—Instrumentation Laboratory, Jarrell-Ash, Perkin-Elmer, and Techtron. Addresses will be found at the end of this appendix.

Even the cheapest instruments of these manufacturers possess amenities over and above basic requirements that add to speed, convenience, and safety in operation. Monochromators have ruled gratings and either reflection optics or zoom lenses in the optical train. Burners are universally of the premix type, with burner heads easily changed. Readouts are four-digit displays, with connections for stripchart recorders. The illuminating system is single-beam and operates on alternating current.

Amenities, not always included in the basic instruments but easily added as options, include flames ignited electrically and in the proper sequence, the zero set pushbutton, signal integration, scale expansion, and the capability of switching to flame emission. Furnace atomization, with the obligatory deuterium background correction, can be added as an option. A turret to hold several hollow-cathode lamps in standby condition reduces waiting time for temperature equilibrium to be reached, thus obviating the single most objectionable feature of single-beam illumination.

The more elaborate outfits are double-beam, with larger monochromators and fitted with two gratings, to cover the visible and ultraviolet regions. Slits are continuously adjustable, and the grating and its scale can be changed either manually or by electric drive. Electronic controls include automatic calibration, automatic rectification of the working curve, and greater expansion of the readout scale. Readout systems provide connections for meter, recorder, printer-sequencer or teletypewriter. Signals can be presented in either normal or peak-holding display.

Two of the manufacturers (Instrumentation Laboratory and Jarrell-Ash) offer instruments consisting of two complete double-beam spectrophotometers, which can be used in the normal way to determine two elements simultaneously or with one of the channels as an internal control, similar to the procedure commonly practiced in emission spectroscopy.

The cost of complete outfits varies over a wide range. The simplest instruments start at about $4000. Larger, double-beam instruments, with two gratings and more elaborate features, range up to about $15,000. It should be remarked here that the main advantages of the more expensive instruments are convenience and speed of operation. In general, very good work can be done with instruments of the lowest price.

Lamps, both hollow-cathode and high-frequency electrodeless, must be ordered separately, according to the analytical needs of the laboratory. The former cost about $150 each for single-element lamps, slightly higher for multielement lamps. Some saving in the cost of the lamp library can be achieved by ordering multielement lamps of seldom-needed elements. Electrodeless lamps cost about $175 each; their useful lifetime is greater than the equivalent hollow-cathode lamps but they require a separate power supply, which costs about $1000.

Furnace atomizers, either for graphite cuvets or for tantalum strip, cost from about $2000 up, the "up" depending on whether the item contains a temperature programmer (a ramp control) and a deuterium background corrector. Automatic readout accessories, such as digital readouts, autoprinters, and teletypewriters, range up to $5000. Certain relatively minor accessories, such as gases, cuvets, burner heads, and so on, must also be considered.

SOURCES OF COMMERCIAL EQUIPMENT

Instrumentation Laboratory, Inc.
  Jonspin Road, Wilmington, MA 01887
    Full line of instruments, lamps, readout devices.
Jarrell-Ash Co.
  590 Lincoln St., Waltham, MA 02154
    Full line.
Perkin-Elmer Corp., Instrument Division
  Norwalk, CT 06856
    Full line.
Varian-Techtron
  611 Hanson Way, Palo Alto, CA 94303
    Full line.
Spex Industries
  Box 798, Metuchen, NJ 08840
    Monochromators, analytical standards, miscellaneous supplies.
Gencom Division of EMI
  80 Express St., Plainview, NY 11803
    Photomultipliers.
Radio Corporation of America, Phototube Marketing
  New Holland Ave., Bldg. 100, Lancaster, PA 17604
    Photomultipliers.
The Ealing Corporation
  2225 Massachusetts Ave., Cambridge, MA 02140
    Osram and Phillips lamps, filters, lenses, mirrors, beam splitters,
    optical benches.
Hamamatsu Corporation
  120 Wood Ave., Middlesex, NJ 08846
    Photomultipliers and Lamps.
Westinghouse Electric Corp., Electron Tube Division
  Elmira, NY 14902
    Hollow-cathode lamps.
Atomic Spectral Lamps Pty, Ltd.
  23-31 Islington St., Melbourne, Australia
    Hollow-cathode lamps (through Techtron).
    Agent: Aztec Instruments, Inc.
      2 Silverbrook Road, Westport, CT 06883

SAFE PRACTICES IN THE LABORATORY

The following recommendations are made to operators of atomic absorption equipment by the Scientific Apparatus Makers Association:*

EXHAUST SYSTEMS

1. Make sure system is drawing properly by smoke test at mouth of collection hood.
2. Make sure duct casing is installed using fireproof construction. Route ducts away from sprinkler heads.
3. Locate blower as close to the outlet as possible. All joints on the discharge side should be airtight, especially if toxic vapors are being carried.
4. Equip outlet end of system with a backdraft damper and take necessary precautions to keep exhaust outlet away from open windows or extended above the roof of the building.
5. Equip the blower with a pilot light located near the instrument to indicate to the operator when the blower is on.
6. Do not make solder joints in the ducting.

GAS CYLINDERS

1. Fasten all cylinders securely to an immovable bulkhead or a permanent wall.
2. When gas cylinders are stored in confined areas, such as a room, ventilation should be adequate to prevent toxic or explosive accumulations. Move or store gas cylinders only in a vertical position.

*1140 Connecticut Ave., Washington, DC 20036.

3. Locate gas cylinders away from heat or ignition sources, including heat lamps. Cylinders have a pressure relief device that will release contents of cylinder if temperature exceeds 52°C or 125°F.

4. Mark gas cylinders clearly to identify the contents without doubt, according to American National Standard Z48.1-1954.*

5. Use only approved regulators and hose connectors. Do not attempt to refill gas cylinders. Left-hand thread fittings are used for fuel gas tank connections, whereas right-hand thread fittings are used for support gas connections.

6. When the equipment is turned off (for example, in the evening), close the fuel gas cylinder valve tightly at the tank. Bleed the remainder of the line to the atmosphere before the exhaust fan is turned off.

7. Arrange gas hoses where they will not be damaged or stepped on, and where things will not be dropped on them.

GENERAL PRECAUTIONS

1. Never run acetylene at a pressure higher than 15 psi. At pressures above this level acetylene can explode spontaneously.

2. Use brass tubing that contains less than 65% copper (2), galvanized iron tubing, or other tubing that will not react chemically with acetylene. Avoid contact between acetylene gas and silver, liquid mercury, or gaseous chlorine. Periodically check for the presence of acetylene in the laboratory atmosphere, especially near the ceiling.

3. Perform periodic gas leak tests by applying a soap solution to all joints and seals.

4. If a burner that uses a mixture of fuel and oxidant is attached to a waste vessel, provide a liquid channel between the burner and the vessel. The head of liquid in the tube should be greater than the pressure in the burner, with a minimum drop of 5 cm of water. If this is not done, mixtures of fuel and oxidant may be vented to the atmosphere, filling the waste vessel and forming explosive mixtures. Waste vessels used with burners should be made of a material that will not shatter in the event of a backflash in the burner chamber. Glass should be avoided.

5. Take suitable precautions when using the more volatile organic solvents in atomic absorption spectroscopy. Feeding the sample capillary

---

*American National Standard Method of Marking Portable Compressed Gas Containers to Identify the Material Contained.

through a small hole made in a covered sample container is one way of handling this problem.

6. Never view the flame or the hollow-cathode lamps directly without protective eyewear; these lamps emit hazardous ultraviolet radiation. Ordinary safety glasses will in general provide sufficient protection, but additional side shields will ensure a further margin of safety. The safety glasses will also provide mechanical protection for the eyes.

7. When using premix burners with cyanide solutions, check the pH of the liquid trap. It may be acidic from earlier work, which can result in the release of a highly toxic gas.

8. Never leave a flame unattended.

Several other precautions, against conditions hazardous not to the operator but rather to the quality of his results, should be borne in mind. If the laboratory illumination is by fluorescent lamps, there is enough leakage of the mercury spectrum, from the yellow lines down to 366 nm in the near ultraviolet, to affect the detector under certain unusual conditions, such as specular reflection from a metallic surface. Smoking by the operator during a run, if the smoke enters the flame, will affect the determination of sodium, potassium, lithium, calcium and magnesium, and perhaps other elements. Room dust also poses a danger; for example, an electric motor with segmented commutator operating in the room will throw off copper dust into the air.

REFERENCES

1. Texts on the history of science and spectroscopy:

   C. Singer, *A Short History of Scientific Ideas to 1900*, Oxford University Press, New York, 1959.

   E. C. C. Baly, *Spectroscopy*, Vol. I, Longmans Green, New York, 1924, Chapter I.

   E. K. Weise, in *Encyclopedia of Spectroscopy*, G. L. Clark, Ed., Reinhold, New York, 1960, pp. 188–199.

   W. P. D. Wightman, *Growth of Scientific Ideas*, Oliver & Boyd, London, 1950.

   Philip Lenard, *Great Men of Science*, Macmillan, New York, 1933.

2. R. A. Sawyer, *Experimental Spectroscopy*, Dover, New York, 1963, p. 1.

3. W. J. Bisson and W. H. Dennen, *Science*, **135**, 921 (1962). (Discussion of Newton's failure to see the Fraunhofer lines.)

4. W. H. Wollaston, *Phil. Trans.*, **92**, 365 (1802).

5. Joseph Fraunhofer, *Ann. Physik*, **56**, 264 (1817).

6. G. R. Kirchhoff and R. Bunsen, *Phil. Mag.*, (4) **20**, 89 (1861). Translation by H. E. Roscoe from original in *Pogg. Ann.*, **7** (1861).

7. J. Tyndall, *Six Lectures on Light*, D. Appleton, New York, 1898.

8. R. W. Wood, *Phil. Mag.*, **10**, 513 (1905).

9. A. C. G. Mitchell and M. W. Zemansky, *Resonance Radiation and Excited Atoms*, University Press, Cambridge, 1934.

10. T. T. Woodson, *Rev. Sci. Instr.*, **10**, 308 (1939).

11. C. T. J. Alkemade and J. M. W. Milatz, *Appl. Sci. Res.* **4B**, 289 (1955). In English.

12. A. Walsh, *Spectrochim. Acta*, **7**, 108 (1955).

13. A. Walsh, *Anal. Chem.*, **46**, 698A (1974).

14. J. P. Shelton and A. Walsh, *Proc. XVth Congr. IUPAC IV*-50, Lisbon, 1956.

15. B. J. Russell, J. P. Shelton, and A. Walsh, *Spectrochim. Acta*, **8**, 317 (1957).

16. D. J. David, *Analyst*, **83**, 536 (1957).

17. J. E. Allen, *Analyst*, **83**, 433 (1958).

18. H. Lundegardh, *Arch. Kemi Min. Geol.*, **10A**, No. 1 (1928); *Z. Physik*, **66**, 109 (1930); *Landbruks—Hogst. Ann.*, **3**, 49 (1936).

19. H. L. Kahn, *At. Absorpt. Newsl.*, **6**, 51 (1967).

20. W. J. Price, *Proc. Soc. Anal. Chem.*, March 1973, p. 63.

21. F. J. Fernandez and J. D. Korber, *Amer. Lab.*, March 1976.

22. W. T. Elwell, *Proc. Soc. Anal. Chem.*, **10**, 53 (1973).

23. F. A. Jenkins and H. E. White, *Fundamentals of Physical Optics*, McGraw-Hill, New York, 1937, p. 274.

W. Grotrian, *Graphische Darstellung der Spektra*, Springer, Berlin, 1928.

B. Cagnac and J-C Pebay-Peyroula, *Modern Atomic Physics: Fundamental Principles*, Halstead Press, New York, 1975.

G. Hertzberg, *Atomic Spectra and Atomic Structure*, 2nd ed., Dover, New York, 1944.

F. K. Richtmyer and E. H. Kennard, *Introduction to Modern Physics*, McGraw-Hill, New York, 1955.

A. C. Candler, *Atomic Spectra*, Vols. I and II, Cambridge University Press, London, 1937.

24. C. E. Moore, *NBS Circ. 467, Atomic Energy Levels*, U.S. Govt. Printing Office, Washington, D.C.: Vol. I, 1949; Vol. II, 1952; Vol. III, 1958.

25. *International Union of Pure and Applied Chemistry, Nomenclature, Symbols*, etc. Pergamon Press, New York, 1975. Reprinted in *Appl. Spectry.*, **31**, 345 (1977).

26. A. C. G. Mitchell and M. W. Zemansky, *Resonance Radiation Excited Atoms*, University Press, Cambridge, 1961.

27. P. W. J. M. Boumans, *The Theory of Spectrochemical Excitation*, Hilger & Watts, London, 1966, p. 83.

28. H. Margenau and W. W. Watson, *Rev. Mod. Phys.*, **8**, 22 (1936).

29. B. P. Straugham and S. Walker, *Spectroscopy*, Vol. III, Wiley, New York, 1976, p. 251.

30. R. G. Breene, *The Shift and Shape of Spectral Lines*, Pergamon, New York, 1061.

31. P. W. J. M. Boumans, *The Theory of Spectrochemical Excitation*, Hilger and Watts, London, 1966, p. 88–91.

32. G. R. Harrison, *M.I.T. Wavelength Tables*, 2nd ed., Wiley, New York, 1969.

33. A. N. Saidel, W. K. Prokofiev, and S. M. Raiskii, *Tables of Spectrum Lines*, VEB Verlag Technik., Berlin DDR, 1955.

34. *Handbook of Chemistry and Physics*, Chemical Rubber Co., Cleveland, Ohio (annual).

35. W. F. Meggers, C. H. Corliss, and B. F. Scribner, *Tables of Spectral Line Intensities*, Parts I and II, National Bureau of Standards Monograph 32, U.S. Govt. Printing Office, Washington, D.C., 1961.

36. M. Margoshes, *Anal. Chem.*, **39**, 1093 (1967).

37. M. L. Parsons, B. W. Smith, and P. M. McElfrish, *Appl. Spectry.*, **27**, 471 (1973).

38. H. L. Kahn, *J. Chem. Educ.*, **43**, A7 (1966); **43**, A103 (1966).

39. W. J. Price, *Proc. Anal. Chem.*, March 1973, p. 58.

40. C. Veillon, *Handbook of Commercial Scientific Instrumentation*, Vol. I, *Atomic Absorption*, Dekker, New York, 1972.

41. M. Slavin, *Emission Spectrochemical Analysis*, Wiley-Interscience, New York, 1971, p. 90.

42. R. M. Barnes and R. F. Jarrell, in *Analytical Emission Spectroscopy*, Vol. I, E. L. Grove, Ed., Dekker, New York, 1971, p. 209.

43. A. H. Sommer, *Photosensitive Materials*, Wiley, New York, 1968.

44. K. B. Mitchell, *J. Opt. Soc. Amer.*, **51**, 846 (1961).

45. H. M. Crosswhite, G. H. Dieke, and C. S. Legagneur, *J. Opt. Soc. Amer.*, **45**, 270 (1955).

46. B. V. L'vov, *Atomic Absorption Spectrochemical Analysis*, Elsevier, New York, 1970, pp. 40–63.

47. J. E. Allen, *Spectrochim. Acta,* **15**, 800 (1959).

48. J. C. Burger, W. Gillies, and G. K. Yamasaki, Westinghouse Product Memo ETD-6403, Elmira, New York, 1964.

49. W. G. Jones and A. Walsh, *Spectrochim. Acta,* **10**, 249 (1960).
    H. Massmann, *Z. Instrumentenk.*, **71**, 225 (1963).
    L. R. P. Butler and A. Strasheim, *Spectrochim. Acta,* **21**, 1207 (1965).

50. C. Sebens, J. Vollmer, and W. Slavin, *At. Absorpt. Newsl.*, **3**, 165 (1964).

51. D. C. Manning, D. J. Trent, and J. Vollmer, *At. Absorpt. Newsl.,* **4**, 234 (1965).

52. J. R. Brandenberger, *Rev. Sci. Instr.*, **42**, 1535 (1971).

53. R. M. Dagnall, K. C. Thompson, and T. S. West, *Talanta,* **14**, 551 (1967).

54. J. M. Mansfield, M. P. Bratzel, H. O. Norgordon, H. O. Knapp, D. O. Zacha, and J. D. Winefordner, *Spectrochim. Acta,* **23B**, 389 (1968).

55. B. V. L'vov, *Atomic Absorption Spectrochemical Analysis*, Hilger and Watts, London, 1970, pp. 64–67.

56. M. Slavin, *Appl. Spectry.*, **20**, 333 (1966).

57. S. Neumann and M. Kriese, *Spectrochim. Acta,* **29B**, 127 (1974).

58. B. M. Gatehouse and J. B. Willis, *Spectrochim. Acta,* **17**, 710 (1961).

59. J. E. Allen, *Spectrochim. Acta,* **18**, 259 (1962).

60. J. B. Willis, *Nature,* **207**, 715 (1965).

61. J. B. Willis, *Appl. Opt.*, **7**, 1295 (1968).

62. L. de Galen and G. F. Samaey, *Spectrochim. Acta,* **25B**, 245 (1970).

63. J. E. Chester, R. M. Dagnall, and M. R. G. Taylor, *Analyst,* **95**, 702 (1970).

64. J. A. Dean, *Flame Emission and Atomic Absorption Spectrometry*, Vol. II, J. A. Dean and T. C. Rains, Eds., Dekker, New York, 1971, p. 235.

65. R. F. Suddendorf and M. B. Denton, *Appl. Spectry.*, **28**, 814 (1974).

66. R. J. Reynolds and D. S. Lagden, *Analyst,* **96**, 319 (1971).

67. J. A. Dean and W. J. Carnes, *Anal. Chem.*, **34**, 192 (1962).

68. J. H. Gibson, W. E. L. Grossman, and W. D. Cooke, *Anal. Chem.*, **35**, 266 (1963).

69. J. D. Winefordner, C. T. Mansfield, and T. J. Vickers, *Anal. Chem.*, **35**, 611 (1963).

70. D. Burgess, *U.S. Bureau of Mines Bull.* No. 604, 1962.

71. M. D. Amos and J. B. Willis, *Spectrochim. Acta,* **22**, 1325, 2128 (1965).

72. L. R. P. Butler, *J. S. Afr. Inst. Min. Met.* **62**, 786 (1962).

73. T. R. Andrew and P. N. R. Nichols, *Analyst,* **87**, 25 (1962).

74. W. Slavin, *At. Absorpt. Newsl.,* **2**, 1 (1963).

75. E. A. Bolling, *Spectrochim. Acta,* **22**, 425 (1966).

76. G. F. Kirkbright and M. Sargent, *Atomic Absorption and Fluorescence Spectroscopy*, Academic, New York, 1974, p. 232.

77. J. B. Willis, *Spectrochim. Acta,* **23A**, 811 (1967).

78. A. S. King, *Astrophys. J.,* **28**, 300 (1908).

79. B. V. L'vov, *Atomic Absorption Spectrochemical Analysis*, translated by J. H. Dixon, American Elsevier, New York, 1970, pp. 193–252.

80. H. Massmann, *Spectrochim. Acta,* **23B**, 215 (1968).

81. J. Y. Hwang, T. Corum, J. J. Sotera, and H. L. Kahn, A.S.T.M. STP 618, American Society for Testing & Materials, 1976, pp. 43–53.

82. J. P. Matousek, *Amer. Lab.,* **3**(6), 45 (1971).

83. M. D. Amos, P. A. Bennett, K. G. Brodie, P. W. Y. Lung, and J. P. Matousek, *Anal. Chem.,* **43**, 211 (1971).

84. Akbar Montasir and S. R. Crouch, *Anal. Chem.,* **46**, 1817 (1974).

85. R. B. Baird and S. M. Gabrielson, *Appl. Spectry.,* **28**, 273 (1974).

86. D. D. Siemer, R. Woodriff, and B. Watne, *Appl. Spectry.,* **28**, 582 (1974).

87. R. Woodriff, *Appl. Spectry.,* **28**, 413 (1974).

88. S. R. Koirtyohann and G. Wallace, 4th *International Conference on Atomic Spectroscopy,* Toronto, 1973, Paper No. 27.

89. H. Massman, *Flame Emission and Atomic Absorption Spectrometry*, Vol. II, J. A. Dean and T. C. Rains, Eds., Dekker, New York, 1971, p. 95.

90. H. L. Kahn, G. E. Peterson, and W. E. C. Schallis, *At. Absorpt. Newsl.,* **7**, 35 (1968).

91. H. L. Kahn and J. E. Sebestyen, *At. Absorpt. Newsl.,* **9**, 33 (1970).

92. H. T. Delves, *Analyst,* **95**, 431 (1970).

93. H. Lundegardh, *Arkiv. Kemi Min. Geol.,* **10A**, No. 1, 1928; *Die Quantitative Analyse der Elemente*, Gustav Fischer, Jena, 1929; *Z. Physik.,* **66**, 109 (1930); *Die Quantitative Spektral Analyse der Elemente*, Vol. II, Gustav Fisher, Jena, 1934; *Leaf Analysis,* Hilger, London, 1951.

94. E. E. Pickett and S. R. Koirtyohann, *Anal. Chem.,* **41**, No. 14, 28A (1969).

95. R. Herrmann and C. T. J. Alkemade, *Clinical Analysis by Flame Photometry,* translated by P. T. Gilbert, Wiley, New York, 1963.

96. R. Mavrodineanu and H. Boiteux, *Flame Spectroscopy*, Wiley, New York, 1965.

97. T. J. Vickers and J. D. Winefordner, in *Analytical Emission Spectroscopy*, Part II, E. L. Grove, Ed., Dekker, New York, 1972, pp. 255–394.

98. J. D. Winefordner and T. J. Vickers, *Anal. Chem.*, **36**, 161 (1964).

99. J. D. Winefordner and R. A. Staab, *Anal. Chem.*, **36**, 165 (1964).

100. T. J. Vickers and J. D. Winefordner, *Analytical Emission Spectroscopy*, Part II, E. L. Grove, Ed., Dekker, New York, 1972, pp. 357–379.

101. Augusta Syty, in *Flame Emission and Atomic Absorption Spectrometry*, Vol. II, J. A. Dean and T. C. Rains, Eds., Dekker, New York, 1971, pp. 197–233.

102. G. F. Kirkbright and N. Sargent, *Atomic Absorption and Fluorescence Spectroscopy*, Academic, New York, 1974, p. 467.

103. J. D. Winefordner, J. J. Fitzgerald, and N. Omenetto, *Appl. Spectry.*, **29**, 369 (1975).

104. J. A. Dean, in *Flame Emission and Atomic Absorption Spectrometry*, Vol. II, J. A. Dean and T. C. Rains, Eds., Dekker, New York, 1971, p. 235.

105. E. J. Prior, *Mineral Processing*, Elsevier, New York, 1965, p. 634.

106. W. J. Sharwood and M. von Bernewitz, *Bibliography of Literature on Sampling*, U.S. Bureau of Mines Publ. No. 2336, 1950.

107. *Standard Methods of Laboratory Sampling and Analysis of Coal*, ASTM Standards, Part 5, American Society for Testing & Materials, Philadelphia, 1950, p. 583.

108. W. F. Hillebrand and G. E. F. Lundell, *Applied Inorganic Analysis*, Wiley, New York, 1959, p. 809.

109. A. D. Wilson, *Analyst*, **89**, 18 (1964).

110. T. A. Wright, *Sampling and Evaluating Secondary Non-Ferrous Metals*, AIME Tech. Publ. No. 81.

111. *Methods for Emission Spectrochemical Analysis*, American Society for Testing & Materials, Philadelphia (any edition).

112. A. B. Calder, *Anal. Chem.*, **36**, 25A (1964).

113. S. E. Allen, *Chemical Analysis of Ecological Materials*, Halstead Press, New York, 1974.

114. M. B. Jacobs, *Analytical Toxicology of Industrial Inorganic Poisons*, Wiley-Interscience, New York, 1967, p. 33.

115. J. Dolezal, P. Povondra, and Z. Sulcer, *Decomposition Techniques in Inorganic Analysis*, American Elsevier, New York, 1968.

116. B. Bernas, *Anal. Chem.*, **40**, 24 (1968).

117. R. J. Guest and D. R. Macpherson, *Anal. Chim. Acta*, **70**, 233 (1974).

118. Y. Hendel, A. Ehrenthal, and B. Bernas, *At. Absorpt. Newsl.*, **12**, 130 (1973).

119. M. Tomljanovic and Z. Grobenski, *At. Absorpt. Newsl.*, **14**, 52 (1975).

120. T. T. Gorsuch, *Destruction of Organic Matter*, Pergamon, New York, 1970.

121. R. E. Thiers, in *Trace Analysis*, J. H. Yoe and H. J. Koch, Eds., Wiley, New York, 1955.

122. G. F. Smith, *Anal. Chim. Acta,* **8**, 397 (1953).

123. G. Middleton and R. E. Stuckey, *Analyst,* **78**, 532 (1953).

124. H. J. M. Bowen, *Anal. Chem.,* **40**, 969 (1968).

125. S. Fujiwara and H. Narasaki, *Anal. Chem.,* **40**, 2031 (1968).

126. F. C. Wright and J. C. Riner, *At. Absorpt. Newsl.,* **14**, 103 (1975).

127. J. E. Barney and G. P. Haight, *Anal. Chem.,* **27**, 1285 (1955).

128. P. O. Bethge, *Anal. Chim. Acta,* **10**, 317 (1954).

129. L. W. Gamble and W. H. Jones, *Anal. Chem.,* **27**, 1456 (1955).

130. R. L. Miller, L. M. Fraser, and J. D. Winefordner, *Appl. Spectry.,* **25**, 477 (1971).

131. G. H. Morrison and H. Freiser, *Solvent Extraction in Analytical Chemistry*, Wiley, New York, 1962.

132. J. Stary, *The Solvent Extraction of Metal Chelates*, Macmillan, New York, 1964.

133. J. A. Dean and J. H. Lady, *Anal. Chem.,* **27**, 1533 (1955).

134. J. E. Allan, *Spectrochim. Acta,* **17**, 467 (1961).

135. G. F. Kirkbright and N. Sargent, *Atomic Absorption and Fluorescence Spectroscopy,* Academic, New York, 1974, p. 491.

136. A. J. Lemonds and B. E. McClellan, *Anal. Chem.,* **45**, 1455 (1973).

137. H. Malissa and E. Schoeffmann, *Mikrochim. Acta,* **1**, 187 (1955).

138. A. I. Vogel, *Quantitative Inorganic Analysis*, 3rd ed., Wiley, New York, 1961.

139. F. J. Welcher, *Organic Analytical Reagents*, Vols. 1–4, Van Nostrand, New York, 1947–1948.

140. J. J. Flagg, *Organic Reagents used in Gravimetric and Volumetric Analysis*, Wiley-Interscience, New York, 1948.

141. L. Gordon, M. L. Salutsky, and H. H. Willard, *Precipitation from Homogeneous Solutions*, Wiley, New York, 1959.

142. G. E. F. Lundell and J. I. Hoffman, *Outlines of Methods of Chemical Analysis*, Wiley, New York, 1938.

143. R. L. Mitchell, in *Trace Analysis*, J. H. Yoe and H. J. Koch, Eds., Wiley, New York, 1957, p. 398.

144. O. Samuelson, *Ion Exchangers in Analytical Chemistry*, Wiley, New York, 1953.

145. F. C. Nachod and J. Schubert, *Ion Exchange Technology*, Academic, New York, 1956.

146. R. Kunin, *Ion Exchange Resins*, Wiley, New York, 1958.

147. D. G. Beichler, *Anal Chem.,* **37**, 1054 (1965).

148. K. Govindaraju, *Anal. Chem.,* **40**, 24 (1968).

149. J. B. Headridge and A. Sowerbutts, *Lab. Pract.*, **23**, 99 (1974).

150. D. A. Tinsley and A. Iddon, *Talanta*, **21**, 633 (1974).

151. J. J. Topping and W. A. MacCrehan, *Talanta*, **21**, 1281 (1974).

152. J. Korkisch and A. Sorio, *Talanta*, **22**, 273 (1975).

153. J. Korkisch, L. Goedl, and H. Gross, *Talanta*, **22**, 281 (1975).

154. J. Minczewski, in *Trace Characterization, Chemical and Physical*. W. W. Meinke and B. F. Scribner, Eds., U.S. Bureau of Standards Monograph 100, Washington, D.C., 1965, p. 103.

155. B. V. Rollin, *J. Amer. Chem. Soc.*, **62**, 86 (1940).

156. L. B. Rogers, *J. Electrochem. Soc.*, **99**, 267 (1952).

157. S. E. Q. Ashley, *Anal. Chem.*, **22**, 1379 (1950).

158. B. V. L'vov, *Spectrochim. Acta*, **17**, 761 (1961). In English. The first paper appeared (in Russian) in a Russian journal in 1959.

159. R. W. Woodriff and R. W. Stone, *Appl. Opt.*, **7**, 1337 (1968).

160. R. W. Woodriff and G. Ramelow, *Spectrochim. Acta*, **23B**, 665, (1968).

161. R. W. Woodriff, R. W. Stone, and A. M. Held, *Appl. Spectry.*, **23**, 408 (1968).

162. W. Slavin, Ottawa Conference on Atomic Spectroscopy, May, 1976.

163. F. J. Langmyhr, Y. Thomassen, and A. Massoumi, *Anal. Chim. Acta*, **68**, 305 (1974).

164. F. J. Langmyhr, J. R. Stuberg, Y. Thomassen, J. E. Hanssen, and J. Dolezal, *Anal. Chim. Acta*, **71**, 35 (1974).

165. F. J. Langmyhr and S. Rasmussen, *Anal. Chim. Acta*, **72**, 79 (1974).

166. F. J. Langmyhr, A. Sundli, and J. Jonsen, *Anal. Chim. Acta*, **73**, 81 (1974).

167. F. J. Langmyhr, Y. Thomassen, and A. Massoumi, *Anal. Chim. Acta*, **67**, 460 (1973).

168. A. D. Wilson, *Analyst*, **89**, 18 (1964).

169. R. Woodriff, *Appl. Spectry.*, **28**, 413 (1974).

170. W. C. Campbell and J. M. Ottaway, *Talanta*, **21**, 837 (1974).

171. D. D. Siemer, R. Woodriff, and B. Watne, *Appl. Spectry.*, **28**, 582 (1974).

172. E. Sabastiani, K. Ohls, and G. Riemer, *Z. Anal. Chem.*, **264**, 105 (1973).

173. D. C. Manning, *At. Absorpt. Newsl.*, **14**, 99 (1975).

174. E. Jackwerth and H. Bernt, *Anal. Chim. Acta*, **74**, 299 (1975) (in German).

175. E. Jackwerth and J. Messerschmidt, *Anal. Chim. Acta*, **87**, 341 (1976) (in German).

176. H. Bernt and E. Jackwerth, *Z. Anal. Chem.*, **283**, 15 (1977) (in German).

177. P. D. Goulden, *At. Absorpt. Newsl.*, **16**, 121 (1977).

178. B. Meddings and H. Kaiser, *At. Absorpt. Newsl.*, **6**, 28, 1967.

179. D. R. Weir and R. P. Kofluk, *At. Absorpt. Newsl.*, **6**, 24, 1967.

180. D. R. Thomarson and W. J. Price, *Analyst*, **96**, 825 (1971).

181. W. J. Price, *Proc. Soc. Anal. Chem.*, March 1973, p. 58.

182. F. J. Fernandez and J. D. Kerber, *Amer. Lab.*, March 1976.

183. J. D. Ingle, *Anal. Chem.*, **46**, 2161 (1974).

184. N. W. Bower and J. D. Ingle, *Anal. Chem.*, **48**, 686 (1976).

185. J. A. Varley and P. Y. Chin, *Analyst*, **95**, 592 (1970).

186. E. D. Truscott, *Anal. Chem.*, **42**, 1657 (1970).

187. R. S. Danchik and D. F. Boltz, *Anal. Letters*, **1**, 901 (1968).

188. R. J. Jacubiec, and R. J. Boltz, *Anal. Chem.*, **41**, 78 (1969).

189. J. A. Goleb, *Anal. Chim. Acta*, **34**, 135 (1966).

190. G. Rossi, *Spectrochim. Acta*, **26B**, 271 (1971).

191. J. A. Wheat, *Appl. Spectry.*, **25**, 328 (1971).

192. J. Yofe, R. Avni, and M. Stiller, *Anal. Chim. Acta*, **28**, 331 (1963).

193. J. B. Willis, *Anal. Chem.*, **34**, 614 (1962).

194. G. K. Billings, *At. Absorpt. Newsl.*, **4**, 357 (1965).

195. R. C. Barras and J. D. Helwig, *Amer. Petrol. Inst. Abstr. Refining Lit.*, May 1963.

196. S. R. Koirtyohann and E. E. Pickett, *Anal. Chem.*, **38**, 585 (1966).

197. W. T. Elwell and J. A. F. Gidley, *Analytical Chemistry (Proc. Feigl Ann. Symp.)*, Elsevier, Amsterdam, 1962, p. 291.

198. D. C. Manning and L. Capacho-Delgado, *Anal. Chim. Acta*, **36**, 312 (1966).

199. S. B. Smith, Jr., J. A. Blasi, and F. J. Feldman, *Anal. Chem.*, **40**, 1525 (1968).

200. F. J. Feldman, J. A. Blasi, and S. B. Smith, Jr., *Anal. Chem.*, **41**, 1095 (1969).

201. M. L. Parsons, B. W. Smith, and P. M. McElfrish, *Appl. Spectry.*, **27**, 471 (1973).

202. W. F. Meggers, C. H. Corliss, and B. F. Scribner, *Tables of Spectral Line Intensities; Part I, Arranged by Elements; Part II, Arranged by Wavelength;* U.S. National Bureau of Standards, Washington, D.C., 1975.

203. R. W. B. Pearse and A. G. Gaydon, *Identification of Molecular Spectra*, 3rd ed., Chapman & Hall, London, 1963.

204. *Nomenclature, Symbols, Units and their Usage in Spectrochemical Analysis. III. Analytical Flame Spectroscopy and Associated Non-flame Procedures.* International Union of Pure and Applied Chemistry 45, 1976, pp. 105–123.

205. B. M. Gatehouse and J. B. Willis, *Spectrochim. Acta*, **17**, 710 (1961).

206. J. E. Allen, *Spectrochim. Acta*, **18**, 259 (1962).

207. S. R. Koirtyohann, 18th Mid-American Symp. Spectroscopy, Chicago, 1967.

208. W. Slavin, *Atomic Absorption Spectroscopy*, Wiley-Interscience, New York, 1968, pp. 60–61.

209. S. Slavin, W. B. Barnett, and H. L. Kahn, *At. Absorpt. Newsl.*, **11**, 37 (1972).

210. *Analytical Methods for Atomic Absorption Spectrophotometry*, Perkin-Elmer, Norwalk, CT, 1973.

211. *Methods for Emission Spectrochemical Analysis*, Part 41, Amer. Soc. Testing & Materials, Philadelphia, 1975, p. 165.

212. M. H. Quenouille, *Introductory Statistics*, Butterworths, London, 1950.

213. W. J. Youdon, *Statistical Methods for Chemists*, Wiley, New York, 1951.

214. W. G. Cochran, *Sampling Techniques*, Wiley, New York, 1963.

215. P. R. Bevington, *Data Reduction and Error Analysis for the Physical Sciences*, McGraw-Hill, New York, 1969.

216. A. G. Worthing and J. Geffner, *Treatment of Experimental Data*, Wiley, New York, 1948.

217. H. F. Rainsford, *Survey Adjustments and Least Squares*, Unger, New York, 1950, p. 207.

218. J. Sherman, "Statistical Analysis," in *Physical Methods in Chemical Analysis*, Vol. II, W. G. Berl, Ed., Academic, New York, 1951, pp. 508–589.

219. F. Twyman and G. F. Lothian, *Proc. Phys. Soc. (London)*, **45**, 643 (1933).

220. N. T. Gridgeman, *Anal. Chem.*, **24**, 445 (1952).

221. R. K. Skagerboe, in *Flame Emission and Atomic Absorption Spectrometry*, Vol. I, J. A. Dean and T. C. Rains, Eds., Dekker, New York, 1969, p. 395.

222. W. Slavin, *Appl. Spectry.*, **19**, 32 (1965).

223. W. A. Hareland, E. R. Ebersole, and T. P. Ramachandran, *Anal. Chem.*, **44**, 520 (1972).

224. G. F. Kirkbright, A. P. Rao, and T. S. West, *Spectry. Letters*, **2**, 69 (1969).

225. H. N. Johnson, G. F. Kirkbright, and R. T. Whitehouse, *Anal. Chem.*, **45**, 1603 (1973).

226. G. F. Kirkbright and H. N. Johnson, *Talanta*, **20**, 433 (1973).

227. G. F. Kirkbright and M. Sargent, *Atomic Absorption and Fluorescence Spectroscopy*, Academic, New York, 1974.

228. J. A. Dean and T. C. Rains, Eds., *Flame Emission and Atomic Absorption Spectrometry*, Vol. III, *Elements and Matrices*, Dekker, New York, 1975.

229. T. C. Rains, *Am. Soc. Testing Materials*, STP 564 (1974).

230. T. S. West, *Analyst*, **99**, 886 (1974).

231. G. M. Hieftje, T. R. Copeland, and D. R. de Olivares, *Anal. Chem.*, **48**, 142R (1976).

232. J. B. Willis, *Endeavour*, **32**, 106 (1973).

233. J. B. Dawson, *Proc. Soc. Anal. Chem.*, **10**, 54 (1973).

234. J. G. Reinhold, *Clin. Chem.*, **21**, 476 (1975).

235. M. B. Jacobs, *Analytical Toxicology of Industrial Poisons*, Wiley-Interscience, New York, 1967.

236. D. J. Lisk, *Science*, **184**, 1173 (1974).

237. I. Sunshine, *Anal. Chem.*, **47**, 212A (1975).

238. W. J. Price, *Metals Mater.*, **8**, 485 (1974).

239. G. Törg, *Talanta*, **19**, 1489 (1972).

240. M. Kubata, D. W. Golightly, and R. Mavrodineanu, *Appl. Spectry.*, **30**, 56 (176).

241. S. Sprague and W. Slavin, *At. Absorpt. Newsl.*, **2**, 20 (1963).

242. D. G. Mitchell, K. W. Jackson, and K. M. Aldous, *Anal. Chem.*, **45**, 1215A (1973).

243. T. T. Gorsuch, *Analyst*, **84**, 135 (1959).

244. P. L. Boar and L. K. Ingram, *Analyst*, **95**, 124 (1970).

245. R. Smith, C. M. Stafford, and J. D. Winefordner, *Can. J. Spectry.*, **14**, 2 (1969).

246. F. S. Chuang and J. D. Winefordner, *Appl. Spectry.*, **28**, 215 (1974).

247. *Standard Methods of Test for Lead in Gasoline by Atomic Absorption Spectroscopy*, American Society for Testing and Materials, D 3237-73, 1973, p. 1180.

248. M. Kashiki, S. Yamazoe, and S. Oshima, *Anal. Chim. Acta*, **53**, 95 (1971).

249. P. Kivalo, A. Visapaa, and R. Backman, *Anal. Chem.*, **46**, 1814 (1974).

250. F. J. M. J. Maessen and F. D. Posma, *Anal. Chem.*, **46**, 1445 (1974).

251. M. Glenn, J. Savory, L. Hart, T. Glenn, and J. Winefordner, *Anal. Chim. Acta*, **57**, 263 (1971).

252. I. W. F. Davidson and W. L. Secrest, *Anal. Chem.*, **44**, 1808 (1972).

253. M. Glenn and J. Savory, *Anal. Chem.*, **45**, 203 (1973).

254. C. Fuchs, M. Brasche, K. Paschen, H. Nordbeck, E. Quellhorst, and U. Peek, *Clin. Chim. Acta*, **52**, 71 (1974).

255. J. A. Fiorino, J. W. Jones, and S. G. Capar, *Anal. Chem.*, **48**, 120 (1976).

256. A. A. Cernik and M. H. P. Sayers, *Brit. J. Ind. Medicine*, **28**, 392 (1971).

257. M. M. Josselow and J. B. Bogden, *At. Absorpt. Newsl.*, **11**, 99 (1972).

258. J. Y. Hwang, P. A. Ullucci, S. B. Smith, Jr., and A. L. Malenfant, *Anal. Chem.*, **43**, 1319 (1971).

259. N. P. Kubisik, M. T. Volosin, and M. H. Murray, *Clin. Chem.*, **18**, 410 (1972).

260. E. Norval and L. R. P. Butler, *Anal. Chim. Acta*, **58**, 47 (1972).

261. J. P. Matousek and B. J. Stevens, *Clin. Chem.*, **17**, 363 (1971).

262. F. J. Langmyhr and P. E. Paus, *Anal. Chim. Acta*, **50**, 515 (1970).

263. J. Korkisch and H. Gross, *Talanta*, **21**, 1025 (1974).

264. K. E. Burke, *Analyst*, **97**, 19 (1972).

265. K. H. Koenig and P. Neumann, *Anal. Chim. Acta*, **65**, 210 (1973).

266. T. Ishizuka and H. Sunahara, *Anal. Chim. Acta*, **66**, 343 (1973).

267. J. C. Van Loon, J. H. Galbraith, and H. M. Aarden, *Analyst*, **96**, 47 (1971).

268. R. W. Cattrall and S. J. E. Slater, *Anal. Chim. Acta*, **56**, 143 (1971).

269. D. C. Manning, *At. Absorpt. Newsl.*, **5**, 63 (1966).

270. D. R. Thomerson and W. J. Price, *Anal. Chim. Acta*, **72**, 188 (1974).

271. F. J. Langmyhr and various co-workers, *Anal. Chim. Acta*, **72**, 79 (1974); **71**, 35 (1974); **73**, 81 (1974); **69**, 267 (1974); **67**, 460 (1973).

272. D. A. Segar and J. G. Gonzalez, *Anal. Chim. Acta*, **58**, 7 (1972).

273. R. D. Ediger, *At. Absorpt. Newsl.*, **12**, 151 (1973).

274. H. G. Griffin and M. B. Hocking, *J. Chem. Ed.*, **51**, A289 (1974).

275. F. D. Pierce, T. C. Lamoreaux, H. R. Brown, and R. S. Fraser, *Appl. Spectry.*, **30**, 38 (1975).

276. R. Woodriff and J. F. Lech, *Anal. Chem.*, **44**, 1323 (1972).

277. W. R. Hatch and W. L. Ott, *Anal. Chem.*, **40**(14), 2085 (1968).

278. S. R. Koirtyohann and M. Khalil, *Anal. Chem.*, **48**, 136 (1976).

279. S. Chilov, *Talanta*, **22**, 205 (1975).

280. A. M. Ure, *Anal. Chim. Acta*, **76**, 1 (1975).

281. H. R. Jones, *Mercury Pollution Control*, Noyes Data Corp., Park Ridge, NJ, 1971.

282. R. Hartung and B. D. Dinman, *Environmental Mercury Contamination*, Ann Arbor Science Publ., Ann Arbor, MI, 1972.

283. F. M. D'Itri, *The Environmental Mercury Problem*, Chemical Rubber Co., Cleveland, OH, 1972.

284. J. P. Macquet and T. Theophanides, *At. Absorpt. Newsl.*, **14**, 23 (1975).

285. B. Krinitz and V. Franco, *J. Assoc. Off. Anal. Chem.*, **56**, 869 (1973); B. Krinitz, *ibid.*, **57**, 966 (1974); B. Krinitz and W. Holak, *ibid.*, **59**, 158 (1976).

286. H. L. Kahn, F. J. Fernandez, and S. Slavin, *At. Absorpt. Newsl.*, **11**, 42 (1972).

287. O. K. Galle and L. R. Hathaway, *Appl. Spectry.*, **29**, 518 (1975).

288. A. Wollin, *At. Absorpt. Newsl.*, **9**, 43 (1970).

289. R. Dunk, R. A. Mostyn, and H. C. Hoare, *At. Absorpt. Newsl.*, **8**, 79 (1969).

290. G. F. Kirkbright and H. N. Johnson, *Talanta*, **20**, 433 (1973).

291. A. K. Babko and V. F. Shkaravskii, *Russ. J. Inorg. Chem.*, **7**, 809 (1962); **6**, 1068 (1961).

292. G. D. Renshaw, *At. Absorpt. Newsl.*, **12**, 156 (1973).

293. J. A. Goleb and C. R. Minkiff, *Appl. Spectry.*, **28**, 382 (1974).

294. Ibid. *Appl. Spectry.*, **29**, 44 (1975).

295. J. F. Lech, D. Siemer, and R. Woodriff, *Sci. Technol.*, **8**, 840 (1974).

296. D. D. Siemer and R. Woodriff, *Spectrochim. Acta*, **29B**, 269 (1974); **28B**, 469 (1973).

INDEX